Best wishes.

The Outer Hebrides

The shaping of the islands

Stewart Angus

The White  Horse Press

First published 1997 by The White Horse Press
    10 High Street Knapwell, Cambridge CB3 8NR, UK
    1 Strond, Isle of Harris HS5 3UD, UK

Set in 10 on 12 point Garamond
Printed in England by Biddles Ltd, King's Lynn

A catalogue record for this book is available from the British Library

ISBN 1-874267-33-2

This book is for my Mother,
Marie Angus

CONTENTS

ACKNOWLEDGEMENTS

No work this complex can be contemplated by one individual unaided, and I have been most fortunate in the calibre of those friends and colleagues who have given generously of their time and expertise to assist me in my task.

The Geology section owes much to my colleague Clive Mitchell, who checked a previous draft, and to another colleague, Nicholas Odling, who checked the revised version.

The glacialogist Dr Jost von Weymarn was a great inspiration in the field and in discussion, and my colleague Dr John Gordon checked two drafts of the glaciation chapter. During the 1970s, discussions with Jost von Weymarn taught me much about the islands and glaciation in particular

The coastal chapters were checked by Professor William Ritchie, who must be the ultimate authority on the landforms of the coastline of north-west Scotland, and on machair in particular. Chapter Six was also checked by my colleagues John Love and David Maclennan, with additional advice provided by another colleague, Mary Harman.

The climate section was checked by Dr Gordon Hudson of the Macaulay Land Use Research Institute, while his colleague Willie Towers, who was always generous with advice when we both worked in Golspie, kindly checked the soil chapter.

Professor James Caird was of immense help in respect of old maps.

My colleague in the Joint Nature Conservation Committee, Dr Pat Doody, kindly supplied the coastal statistics.

The entire text was checked by three academic referees. Both my colleague Dr Des Thompson and my old friend Dr Chris Ferreira, have not only been responsible for considerable enhancements in the book by commenting upon the penultimate draft, but have encouraged, nagged and inspired me throughout this project. Professor Michael Usher and my Director, Dr Jeff Watson, also commented upon the text, the former in some detail.

Bill Lawson, a man with a uniquely broad knowledge of the Western Isles, was good enough to referee my references to the human heritage, and he and his wife Chris have been of considerable assistance as this work gathered momentum, giving me generous access to their outstanding library and their combined memory banks. On the subject of libraries, I marvel at the patience of SNH Librarian Alwyn Coupe who, with her team, supplied me with the most obscure information, while the staff of Western Isles Libraries, and David Fowler in particular, performed services well beyond the call of duty when I needed information quickly. Gill Maclean (Howmore) kindly gave me access to her unpublished research into the history of drainage in the Uists, which will be covered in greater detail in the next volume in the series.

viii

Comprehension and readability have been checked by my sister Shona Angus (never mind the sibling relationship – as a Press Officer for SNH she seems extremely well qualified for this task), while my friend Elizabeth Pearson performed a similar task from the point of view of a 'Hebridophile'.

I have been most fortunate in the calibre of my consultees. If, despite their efforts on my behalf, any mistakes remain, then the responsibility is entirely mine.

To Dr Finlay Macleod, John Murray and David Mackay, who funded and encouraged me when I was still a student, and stimulated a body of work of which this book forms the first published instalment, I owe a great debt of gratitude. It is both a pleasure and an honour to reproduce one of John's poems at the head of Chapter Six, and I am additionally grateful to him for permission to print it here.

My first venture into machair research was the idea of my then line manager Sandy Maclennan who, together with colleagues Mary Elliott, Alan McKirdy, Peter Tilbrook and Jeff Watson has been a constant source of encouragement in this field, as has Dr Martin Kent of the University of Plymouth.

Others who have helped in various ways, through discussion and provision of information, include: Ian Armit (University of Edinburgh), Steve Atkins (SNH), John Baxter (SNH), Reg Brown (Holm), Ray Burnett (Benbecula), Gail Churchill (SNH), Trevor Cowie (National Museums of Scotland), Jim Crawford (Garynahine), Gordon Cumming (Coastal Access Project), Joan Cumming (SNH), Andrew Currie (Skye), Tom Curtis (National Parks & Wildlife Service, Ireland), Tom Dargie (Dornoch), Kathy Duncan (SNH), Fiona Everingham (peatland consultant), Johanne Ferguson (SNH), Professor Dave Gilberston (University of Aberystwyth), Ray Goodge (Uig), Norman Johnson (Lochmaddy), Carol Knott (W. Isles Tourism Development Project), George Lees (SNH), Richard Lindsay (peatland consultant), Donald Macaulay (Bernera), John Macaulay (Flodabay), Annie Macdonald (North Uist), Donnie Macdonald (Rodel), Iain Gordon Macdonald (Ness), Roddy Macdonald (WIC), Murdo Mackenzie (S. Uist), Susan Maclennan (W. Isles Tourism Development Project), Hector Macneill (Tangusdale), Jon Merritt (British Geological Survey), John Miles (Scottish Office), Donald Omand (University of Aberdeen), Donald Paterson (University of Aberdeen), Frank Rennie (Lews Castle College), Ian Stephen (Stornoway), Jill Strawbridge (Minch Project), Albert Salman (European Union for Coastal Conservation), Tony Scherr (Borve, Harris), James Simpson (National Museums of Scotland), Katalin Svehla (London), Margaret Wilkes (National Library of Scotland) and Philip L. Woodworth (Permanent Service for Mean Sea Level). I am most grateful to the above (some of whom will feature more prominently in acknowledgements in the next book) for their encouragement and generosity.

Finally, my thanks to publishers Andrew and Alison Johnson. It is gratifying to have one's book published by friends, and only writers will understand when I say that they have produced exactly the book I would wish for.

1

INTRODUCTION

The water is seldom very deep, so that in many places the ground may be seen to a great distance from the land. Agreeably to this deposition, very extensive tracts of land are laid bare, when the tide retires from the shores. These tracts are of frequent occurrence along the whole extent of the range. Some of the islands are separated by sands of this kind, a communication being established between them at low water, so that in fact they are islands only when the tide is up. Benbecula is, in this manner, separated from North Uist on the one side, and South Uist on the other. The other two sounds between the large islands, namely the Sound of Harris, and the Sound of Barray, although never left dry, are, for the most part, shallow, especially the latter; and, for this reason, as much as any other, the whole range has received the general designation of Long Island. (MacGillivray 1828)₅.

There can be few places in Europe where the relationship between rocks, climate, wildlife and human history is clearer than in the Outer Hebrides, or

Stewart Angus, September 1983

Figure 1.1. North Ford at high tide

where people live more intimately with nature. Ancient rocks appear at every turn; among these rocks soils cling to the sheltered crannies; human activity clings to the islands' edges, seeking the ameliorating influence of the sea. A recent strategic study of tourism in the Western Isles, revealed that landscape and scenery was the most important aspect of the islands in attracting visitors, by a very wide margin, while local community appraisals have confirmed that the people who live in the islands care very deeply about their environment. Having been born and brought up in the Western Isles, and having worked in natural heritage conservation in the islands for over eleven years, I have always been aware of the islanders' love for the land and sea around them, and the numerous questions I have been asked as a professional in the field testify to a deep interest and keen observation among islanders. This book attempts to answer at least some of these questions in a structured context.

In 1976, I was commissioned by the Western Isles Bilingual Education Project and the allied Community Education Project to produce a range of teaching information on the shore for island schools. The project directors, John Murray (BEP) and David Mackay (CEP), co-ordinated by Dr Finlay Macleod, funded me to carry out one of the most enjoyable tasks I have ever been asked to perform, a project which lasted into 1983. While it had been the intention of the sponsors to produce some of the output in book form, the report was never published. This book is very, very different from anything envisaged at that stage, and is aimed more widely than the formal education sector, but some of the ideas of these early days have persisted.

The Outer Hebrides: the shaping of the islands deals with the physical aspects of the natural history of the Outer Hebrides, and is designed as the first of three volumes devoted to the natural heritage of these islands; the subsequent volumes will deal with the terrestrial and marine habitats respectively. Though the three are designed to work as a series, each volume is also designed to stand alone. The aim in the series has been to produce a regional natural history which is both comprehensive and comprehensible. Considerable efforts have been devoted to ensuring that the text is both accurate and understandable. I have a particular obsession with ensuring that all the angles are covered, other than those which really are too esoteric to merit inclusion, which is part of the reason that this work has taken nearly twenty years to complete. In respect of machair management – be it for agriculture or conservation – and the use of peat by local people, much seems to be left unsaid. The intention is to deal with these aspects more fully in the next volume, while the influence of the sea is best left for the final volume in the series, instead of trying to include every conceivable aspect of the physical background here.

Like many others of my generation, I was inspired as I grew up by the Collins New Naturalist *Highlands and Islands* by Fraser Darling and Morton Boyd[3], made accessible to all by the issue of a Fontana paperback edition. Darling and Boyd, however, were selective in their writings: of necessity in any single volume some geographical areas are covered in much more detail than others, as some aspects of the environment receive more coverage than others reflecting, no doubt, the authors' interests. Even this fine work was eclipsed by W.H. (Bill) Murray's magnificent *The Islands of Western Scotland*[6], for me the finest book ever written on the Hebrides, now unfortunately out of print. Though our knowledge of the islands has advanced significantly since these great works were published, they have never been surpassed.

In 1977 and again in 1981, the Royal Society of Edinburgh, Scotland's premier scientific organisation, held conferences on the Outer and Inner Hebrides respectively, in association with the Nature Conservancy Council. Such symposia are always selective in their coverage, as their function is to bring academics together, and the resulting publications are aimed squarely at professionals. During the meeting on the Inner Hebrides, a speaker approaching the end of his presentation embarked on an impassioned account of the natural heritage of the islands and, as he spoke, the hall filled with the growing sound of Mendelssohn's *Fingals' Cave*, as he expressed his desire as a conservationist that the islands would go on to inspire in the future as they had in the past. I heard only one criticism of this dramatic departure from scientific tradition – perhaps scientists are as sentimental as everybody else when it comes to the crunch – for scientists to deny that there is something inspirational in the stunning setting of the Outer Hebridean environment is to lose out on something beyond value.

The Hebrides Overture was about the Inner Hebrides, but the outer islands have inspired numerous Gaelic songs and *bardachd* (poetry), the poetry of Ian Stephen, John Murray and Kenneth White, and the paintings and sculptures of artists of the calibre of Iain Brady.

The island landscapes and seascapes which have inspired the visual and written Arts have also inspired scientists. The rocks are not, as will be pointed out, the oldest in the world, but the Outer Hebrides are probably a bit easier to study than Greenland. The gneisses which seem so grey and uniform are in fact extremely varied, with subtle differences enabling geologists to decode the history of the evolution of the Earth's crust. It would be difficult to overestimate the importance of the rocks of the Outer Hebrides for research.

More recent glaciations still seem to pose more questions than answers, and there is nothing scientists enjoy more than a good puzzle. The coast, which makes up one-sixth of the length of Scotland's coastline, seems almost infinitely varied, and its history and development offer great scope for studies in physical geography. When you link these features to the stunning landscapes and the fascinating human history which is now increasingly recognised by scientists as a vital component of their study, it is difficult to imagine a more inspiring geographical area to study ... but I admit to a certain bias here.

The climate can be extremely pleasant when most mainland scientists find they have time to spare for a Hebridean interlude, though there is a danger of missing out some of the effects of climatic extremes, some of which are of fundamental importance in the way people and environment co-exist.

The habitat for which the Outer Hebrides are famous is the machair – perhaps the only major habitat to be known by its Gaelic name. Restricted globally to the north-west of Scotland and the north-west of Ireland – coincidentally (or not?) the Gaelic-speaking areas of both countries – machair embraces all of the elements of this book, but one should not forget that its history is inextricably linked with human history from the earliest stages of both in the islands. I take some pride in asserting that my use of Gaelic poetry as evidence for a scientific case may be a first. Machair is now my special interest, and a subject to which I will one day devote a whole book.

THE ISLAND LANDSCAPE

It could be argued that there is as much seascape as landscape in the Outer Hebrides. The human association with the coastline is almost total with respect to settlement, and the coast is probably much more important than the inland areas for tourism. The form of the landscape is determined by a combination of geology, glaciation, climate, wildlife and human influence.

In the north of Lewis, the great plateau of low-lying peatland extends southwards from Ness to a line connecting Loch Erisort with East Loch Roag. Most of the population is concentrated around Broad Bay, where the land is kinder, and the chief town of Stornoway is within easy commuting distance. Beyond lie the hills of Uig and Pairc, the former gaunt and rocky, the latter more rounded and vegetated. The 'mountains' of Harris are really hills, but are no less dramatic, as they rise so steeply from sea level. These hills display much evidence of glaciation. What greater contrast can there be than that between the west and east coasts of South Harris, the former green and beach-girt, the latter an inhospitable, rocky landscape which has its own austere beauty? Across the rock-strewn Sound of Harris, with its treacherous currents, North Uist lies lower than the wet, peaty platform of northern Lewis, so much so that at High Water, the sea invades the 'inland' lochs. Benbecula, part of the Parish of South Uist, more closely resembles North Uist. South Uist is another island of contrasts, with its rocky east coast separated from the superb western machairs by a ridge of high hills. Barra, now linked to neighbouring Vatersay by a causeway, has machair

Stewart Angus, October 1987

Figure 1.2. Aerial view of Loch Druidibeg and Stilligarry machair over the summit of Hecla, South Uist, showing the contrast between moor and machair. Howmore River is visible on the left, Loch an Eilean (with the island) connected to Loch Druidibeg by a drainage channel, and the cultivated machair fields in the distance.

and beaches on its west and north coasts, where the Tràigh Mhor forms the only beach landing strip in Britain for a scheduled air service. The east and south coasts of Barra seem more hospitable than the east of Harris or the Uists, while Castlebay surpasses Stornoway in the grandeur of its harbour setting.

For many, the coastline is unsurpassed for scenic value. From the plunging cliffs of Conachair on St Kilda to the most sheltered bay at the head of the longest sea loch, lies an infinite range of variation. Tricks of the light and the state of the sky and the sea ensure that the scene is constantly changing. The stunning turquoise hues of sand seen through the sunlit Atlantic waters seem unreal, as though photographically enhanced. When this is part of a view embracing a flower-studded machair, pristine, sweeping beaches, sun-rippled lochs, and a panorama of sunlit hills, with background sounds supplied by a thousand birds, where else could anyone wish to be?

The offshore islands are arguably even more special. Each group has its own atmosphere, which almost always owes as much to the past human inhabitants as to the physical scene. St Kilda seems to have spawned more purple prose than any other part of the Earth, and almost begs to disappoint the first-time visitor – after all – what island can live up to hype like that? St Kilda does, and more, and the feeling of awe does not diminish with repeated visits – small wonder that some people become obsessed with it. Truly, this is a World Heritage Site.

NAMING OF PARTS

There is a famous description of the Burren in western Ireland, a huge expanse of bare limestone in County Clare, attributed to one of Cromwell's soldiers, that it had "not enough wood to hang a man, not enough water to drown him, not enough clay to cover his corpse". The Western Isles has its own version of this, attributed to the noted folklorist Alexander Carmichael[2] who described the north end of South Uist thus:

> On the east, between the ragged townlands [of Iochar] and the Minch, lies a moor interspersed with rocks, bogs, and water. Where the land is not rock it is heath, where not heath it is bog, where not bog it is black peaty shallow lake, and where not lake it is a sinuous arm of the sea, winding, coiling, and trailing its snake-like forms into every conceivable shape, and meeting you with all its black slimy mud in the most unexpected places.

This is almost as bad as Ada Goodrich-Freer's "South Uist is surely the most forsaken spot on God's Earth"[4]. Professor John Stuart Blackie[1] was not being entirely serious when he penned the following lines in respect of Benbecula, which earned him a stern rebuke in a letter to the *Inverness Courier* from no less than Alexander Carmichael:

> O, God forsaken, God detested land
> Of bogs and blasts, and moors and mists and rain.
> Where men with ducks, divide the doubtful strand
> And shirts when washed are straightway soiled again!

Inevitably much of the literature of the Western Isles has been contributed by visitors, some of whom were better informed than others. When one thinks of some of the material written today about the islands, it is of some concern that for many areas of study, it is difficult to find local sources of written information. How fortunate we are that the visitors included such people as Carmichael, John Lorne Campbell, and his wife Margaret Fay Shaw. To this day, the oral tradition of the islanders is not given its rightful place, and there is a danger that much material that is of immense value to researchers, as well as for its own sake, will be lost forever if the surviving traditions are not recorded, as less and less is transferred to the succeeding generation.

The very name 'Outer Hebrides' conjures up an image of islands of mystery, remoteness, and awesome beauty. It is an image which attracts enough tourists to put that 'industry' on an economic par with Harris Tweed, yet one which is often decried within the islands, where many prefer the more prosaic 'Western Isles', despite the possibility of confusion with the Inner Hebrides, who believe that they, too, are the 'Western Isles'.

In the second century A.D. Ptolemy, probably using observations made by the Roman Fleet, called the islands *Ebudae*, though it has to be admitted that he was talking of some of the southern Inner Hebrides[6]. The word is similar to the Viking term *Havbredey* (pronounced 'Haubredey') meaning 'Islands on the Edge of the Sea'.

Place-names are a minefield. For the record, I have used 'Western Isles' in this book only as a synonym for the Outer Hebrides: if I mean the Inner Hebrides, I say so, though I have occasionally referred simply to 'Hebrides' when I mean Inner and Outer together. On a more local level, place-names have long caused problems for map-makers, and the growing trend towards the reversal of the anglicisation of many place-names is most welcome, even if it means that names are presently in a state of flux. The names of even large towns and geographical features depend on which edition of map you examine. I am afraid that I have been completely arbitrary in my selection, but I have not used any in such a way as to leave the reader confused about my meaning.

Map surveyors were not often Gaels, and one suspects that a surveyor's increasing impatience, leading, perhaps, to the question "Whose rock is this?" was responsible for a bodach telling him that the name of a rock in Camas Uig (NB0333) was 'Bodha Co-leis-e' (literally, 'whose rock?'), which he duly noted down and transferred to his chart.

USING THE BOOK

The sequence of chapters is carefully structured, each one relying heavily on those which precede it. Some subjects more readily lend themselves to anecdotal asides than others, and I have taken almost as much pleasure in gathering these stories as in the compilation of the scientific content.

The references are given in a standard form for anyone who wishes to follow up my sources through libraries, and much of the historic (and, increasingly, the scientific) material is available for consultation in island libraries. The list is necessarily selective, as I cannot refer by name to every single relevant work.

A book like this without an Index would be of limited value, and this Index is designed for the confirmed information hunter.

I hope that there are readers who enjoy this book as much as I have enjoyed writing it.

2

GEOLOGY

There is no part of Scotland which rivals, in the hideousness of its bareness and desolation, a stretch of miles in extent which lies to the east of the middle hills of Harris. The moor of Rannoch is, in comparison therewith, a well-watered garden. (Heddle 1888)[32].

Apart from the more extensive peatlands of northern Lewis and some of the larger machairs, there is no escaping the rocks of the Outer Hebrides and, in parts of south-west Lewis and Harris, the landscape is so extraordinarily bare that close inspection is necessary to confirm that there is any soil at all. Much of the landscape is gnarled and knobbed, crushed and creviced, seeming to reflect the great age of the island rocks.

At first glance, the geology of the Outer Hebrides appears relatively simple. Certainly one could mention the enigmatic conglomerates of the Stornoway Formation, throw in a few volcanic dykes for good measure, call all the rest 'Lewisian Gneiss' and leave it at that. This generalised view of the rocks of the Outer Hebrides might almost be said to be generally accurate, especially if you just say 'Lewisian' rather than 'Lewisian gneiss' but it is not all that useful in practice: it does not, for example, explain why the gneiss is so variable, even to the untrained eye, or why it is folded in some areas and not in others. Nor does it consider the most fascinating aspect of all – how detailed studies of the gneisses have revealed the history of the early development of the Earth's crust, a story that takes us back to a time when the interior of the Earth was some 2-3 times hotter than at present, and the crust was so different that fragments can still be identified even in rocks which have been subjected to subsequent changes.

The study of the Lewisian rocks is like a Russian doll: the more detail you go into, the more complex it becomes and, to add to this complexity, different geologists have used different terms for the same or similar rocks and geological events. Most of the writing on the geology of the islands has so far been either very general, or has been aimed at other geologists, and the most useful sources of information, including the standard geological maps, are liberally spattered with technical terms.

Of all the subjects covered in this book, geology probably employs more technical terms than any other. There are but three choices in writing on geology: to use the jargon, to employ long definitions, or to risk inaccuracy. I have attempted to reach the best compromise between the first two in the hope of avoiding the third. Most readers will want to refer to geological maps from time to time. As keys to these invariably use technical expressions, the reader will wish explanations of these, but in order to make the text more readable, technical explanations have been consigned to a series of information boxes which may be consulted as required. Non-geologists should now refer to these before reading on, or refer to them as necessary.

The interpretation of Lewisian geology relies heavily on mineralogy, a field which is rather too esoteric to merit detailed coverage here, and those seeking the whole truth, so far as it is known, are referred to the excellent 1:100,000 scale geological maps produced by the British Geological Survey[5] and the accompanying Memoir: *Geology of the Outer Hebrides*[21]. *Outer Hebrides: localities of geological and geomorphological importance* published by the Nature Conservancy Council (NCC)[4], is probably more useful to the amateur, espe-

---

BOX 2.1 – ROCK TYPES

All rocks are divided into three main groups on the basis of their origins: **sedimentary, igneous** and **metamorphic.**

Igneous rocks are usually formed from the solidification of molten rock by cooling (see Box 2.2).

When rocks are eroded, e.g. by abrasion or weathering, the products of the erosion are deposited elsewhere as **sediments.** Sediments include **shingle, sand, mud** and **clay**, and may include material of biological origin such as shell. Eventually pressure from the weight of overlying sediments or the action of chemical agents (including water) may consolidate the sediments to form **sedimentary rocks**, e.g. clay becomes **shale**, sand becomes **sandstone**, mud becomes **mudstone** and shingle becomes **conglomerate.** While conglomerate has pebbles, a **breccia** has angular stones. **Marl** is a calcareous mudstone and **dolomite** is a rock (usually limestone) containing more than 15% magnesium carbonate.

Sedimentary or igneous rocks buried deep in the Earth's crust or close to igneous activity or to earth movements may be subjected to intense heat and pressure. The combination of heat and pressure is not always strong enough to induce melting, but affected rocks may be physically deformed and many of their minerals chemically altered, so that the rocks are said to have been **metamorphosed.** Metamorphic rocks (see Box 2.3) may themselves be altered by subsequent metamorphic events.

cially as a guide to particular localities. A *Geologists' Association Guide*, which will give detailed coverage of particularly interesting localities, is in preparation at the time of writing.

Beginners in geology often become obsessed with the identification of rock specimens, some of them very small, and tend to use geological maps as identification guides. The inexperienced amateur must accept that there is no way he or she is going to be able to identify all the rocks occurring in the Western Isles – even the professionals need to resort to the study of thin sections of rock samples viewed under geological microscopes – and the fact that the map shows one particular rock type occurring in a particular area does not necessarily mean that all the rock in that area is of that type. The best way for an amateur to learn about the rocks of the Western Isles is by working his way through the localities described in the BGS Memoir$_{21}$ or the NCC guide$_4$, starting with the easy rocks like the conglomerates of the Stornoway Formation, and working through the grey gneisses, the granite-migmatite complex, and the Outer Hebrides Thrust. With the possible exception of the anorthosite, the South Harris Igneous Complex is just that – complex – and should be left till last.

Many of the best exposures are in quarries or road cuttings, and these are often employed by geologists to highlight points of interest. By their very nature, however, these are subject to change, and it is always possible that excellent teaching exposures may disappear through road works or new quarry workings. The wonderful exposures of Scourie Dykes in Howmore Quarry, for example, which are described below, were recently buried beneath thousands of tonnes of boulders as part of the landscaping of a new road (Fig. 2.2). The Lewisian rocks of the north-west of Scotland are part of a much bigger picture, and are best looked at in a wider context rather than as isolated exposures; they yield clues about the relationship between the Earth's crust and its interior at a time when the surface of the planet was still relatively hot and mobile (though not to the extent that it was molten).

Identification of the processes is more important and more interesting than the assignation of a technical name to a rock sample. This Chapter deals with the processes and geographical examples, and supplies the technical labels for those who require them. While Box 2.5 requires a nodding acquaintance with the contents of the rather less gripping Boxes 2.1–2.4, all is explained by plate tectonics (Box 2.5) – at last, it all makes sense – and the vital importance of mineralogy becomes easier to appreciate.

The gneisses, schists and granites of the Outer Hebrides are collectively described as 'Lewisian', and belong to the longest geological period, the Precambrian, which spans the time from the formation of the Earth's crust – probably about 4500 million years ago (Ma) – to 570 Ma. Indeed, the Precambrian spans more than the rest of geological time put together. To put this in a more familiar context: if the age of the crust is represented as a 24-hour day, the earliest rocks of the Outer Hebrides have been formed by around 8am, while Rockall and St Kilda were formed around two minutes to midnight.

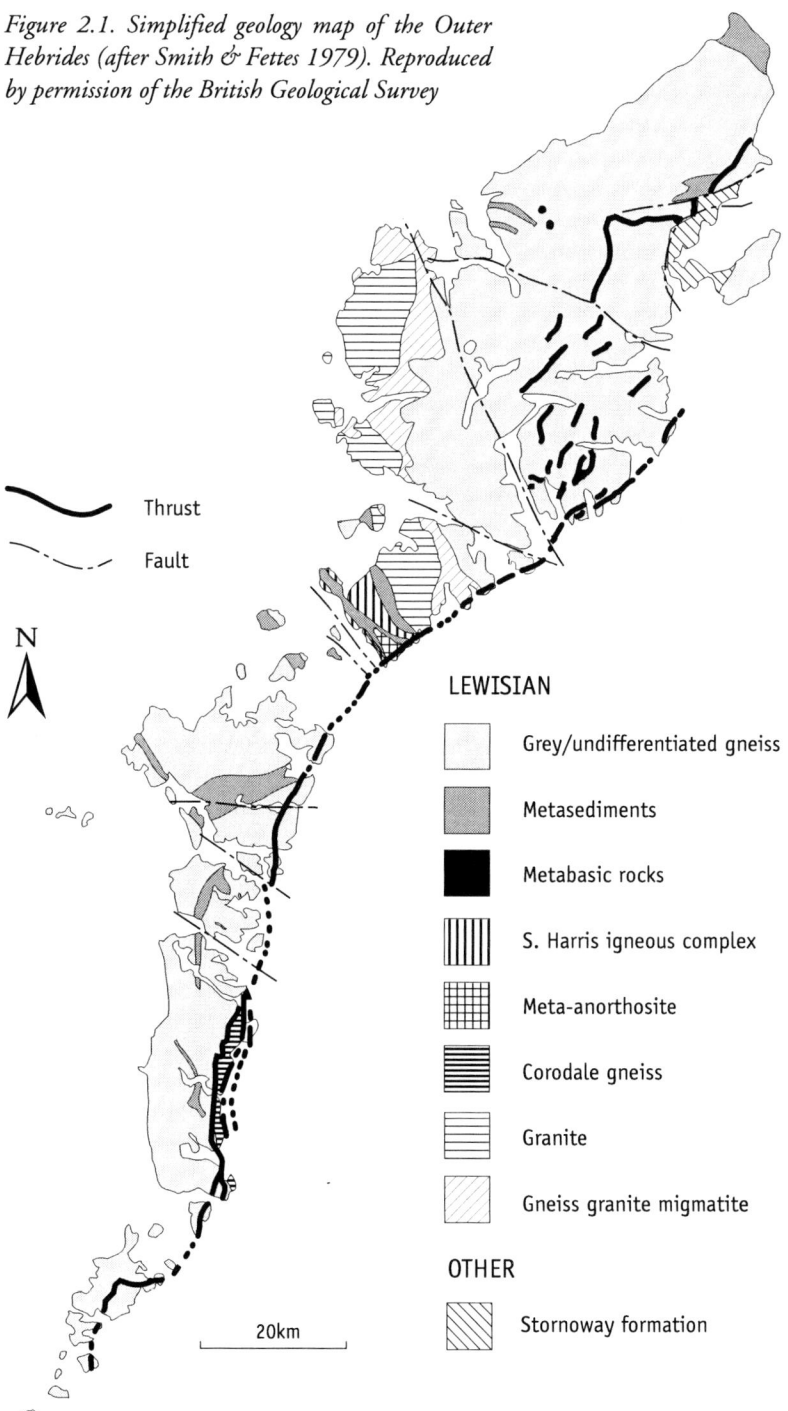

Figure 2.1. Simplified geology map of the Outer Hebrides (after Smith & Fettes 1979). Reproduced by permission of the British Geological Survey

Thrust

Fault

N

LEWISIAN

Grey/undifferentiated gneiss

Metasediments

Metabasic rocks

S. Harris igneous complex

Meta-anorthosite

Corodale gneiss

Granite

Gneiss granite migmatite

OTHER

Stornoway formation

20km

Figure 2.2. Scourie Dykes in Howmore Quarry, South Uist, a site now obscured by road landscaping. The black vertical streaks on the left of the picture are relatively unaltered, whereas those on the right have been deformed by the Laxfordian metamorphism.

Lewisian rocks crop out over much of the north-west of Scotland, where they form the 'Ancient Foreland'. Further east, this Precambrian basement is covered by more recent rocks – mainly Torridonian and Cambrian sediments. The basement extends south-eastwards on the mainland at least as far as the Great Glen Fault, buried by younger rocks to a depth of up to 16km[30]. With the exception of intrusions and the sediments of the Shiant Islands (pronounced 'Shant') in mid-Minch, the only rocks lying over the Lewisian in the Outer Hebrides are the sediments of the Stornoway Formation. Westwards, this basement extends to the margin of the continental shelf[77] (see Box 2.5).

As the great J.J.H.Teall wrote in 1895[58] "Lewisian gneiss is not a geological formation in the ordinary sense of the word". Gneiss is a generic term for a range of rocks of similar, but far from identical, origin (see Boxes 2.1, 2.2, 2.3). The word 'gneiss', the source of some quite inexcusable puns, is derived from a Slavonic word meaning 'nest', the rock being the 'nest' of the ores sought by the miners of Central Europe in medieval times[33].

The Lewisian rocks are very closely related to those of the Rockall Bank, southern Greenland, and north-east Canada; these areas were all part of the same continental mass – the Laurentian Shield – long before the opening of the North Atlantic [77,21]. Indeed, it is believed that the rocks of the Outer Hebrides were formed in association with earlier continental spreading, of which more below.

Lewisian gneiss of the Western Isles is often described as "the oldest rock in the world". It is always a shame to spoil a good story, but the oldest dated rock in the Outer Hebrides is at Ardivachar, with an age of around 3,000 Ma, making this merely *one of the oldest* rocks in Britain, because even its status as the oldest British rock has recently been superseded by the discovery of older (3300 Ma) rocks around Gruinard Bay in Wester Ross[7] – and even this very old date for the Ardivachar rock has been questioned (see below). The oldest rocks in the world are thought to be the Acasta gneisses of the Slave area of north-western Canada, which are about 3960 Ma old[25] though an age of more than 4100 Ma has been suggested for single grains of zircon in Australia[11].

Though a minerals report of Lewis was prepared for Lord Seaforth by the Rev. J.Headrick in 1800[31], and there are annotations (added at a later date) about minerals on Bald's map of Harris of 1804[1], the first detailed description of the geology of the Western Isles was made by John MacCulloch in a series of letters to his friend Sir Walter Scott[42]. It was MacCulloch who first referred to the rocks as 'Lewisian'. Subsequently the islands were visited by some of the great men of geology, including Murchison and Geikie[52] and perhaps the greatest of all Scottish geologists, Benjamin Peach and John Horne[57]. The first really comprehensive account of the geology of the Outer Hebrides was compiled by T.J.Jehu and R.M.Craig of the Geological Survey[34-38].

Sadly, few tales survive of how the people of the Outer Hebrides reacted to strangers who wandered around chipping rocks all day, but the unpublished diaries of the great Victorian naturalist Harvie-Brown describe a visit he made to 'Strome' (almost certainly Stromay in the Sound of Harris NF9888) with the geologist Professor Heddle, in July 1887. A block they had quarried out was destined to be cut up into "3 handsome table-tops" in Aberdeen, and they hired a boat and crew of three men who sailed and rowed out two miles but "never looked at it or the place, but just turned and went away back. Of course they were not paid". It would have been interesting to hear the crew's description of their treatment. The names of Peach and Horne are still held in awe in parts of western Sutherland, where they made discoveries that changed the science of geology,

---

BOX 2.2 – IGNEOUS ROCKS

All of the igneous rocks of the Outer Hebrides originate from the cooling of **magma** (molten rock). When the magma is extruded as **lava** on the surface of the Earth's crust it is **volcanic. Tuffs** are formed from volcanic ash or dust. Magma on or near the surface cools quickly, forming fine-grained rocks such as basalt, or even glassy rocks. When the magma cools deep in the crust it is termed **plutonic** or **hypabyssal**. The slower the cooling, the larger the crystals. Dolerite (see below) cooling uniformly in sheets forms columns which are usually hexagonal, as seen on the Shiant Islands.

Igneous rocks are classified on the basis of their mineral composition, on a scale ranging from **acid** to **ultrabasic**, a terminology which has nothing to do with the actual acidity or alkalinity of the rocks. Any igneous rock containing large amounts of quartz is **acid**, e.g. **granite**. Quartz-free rocks are **basic** and contain calcium-rich plagioclase feldspars, together with olivine and pyroxene (see Box 2.4). Rocks **intermediate** between basic and acid have plagioclase feldspars in which sodium is more abundant than calcium. Unfortunately, the analysis of plagioclase, so essential to the interpretation of igneous rocks, is a task for skilled geologists, who examine a thin section of the rock under a special geological microscope equipped with a rotating polarising filter. **Ultrabasic** rocks are also quartz-free, and have large amounts of iron and magnesium minerals such as hornblende and biotite, which give the rocks a very dark appearance.

Where a magma cuts across the 'grain' of the rock into which it is **intruded**, it solidifies as a **dyke**, a sheet usually exposed 'end-on' as a narrow band. Where the magma intrudes between rock layers, along lines of weakness, it forms a **sill**.

Basic magmas tend to be relatively free-flowing, thus enabling them to form extensive dyke 'swarms' and sills. Basic magmas cooling near the Earth's surface form fine-grained rocks such as **basalt. Dolerite** has much the same mineral

and there is a memorial to them at the Inchnadamph Hotel. The only memorial to any geologist in the Outer Hebrides is a cairn south of Skigersta in Ness commemorating J.Wilson Dougal.

There are also some wonderful stories from Skye which are of course not entirely relevant but are so good as to warrant repetition here, on the excuse that it may give us a clue as to how geologists were regarded in the islands generally.

The Sgiathanachs were so incensed by MacCulloch's descriptions of them that at the first opportunity, Mr Mackinnon of Coirechatachan took a published portrait of the author to Glasgow, and commissioned a set of chamber pots displaying MacCulloch's features[26]. Another Skye-man was overheard by one of Geikie's friends describing his experience of geologists (in Gaelic): "There

---

composition as basalt, but has cooled more slowly, and is medium-grained. The coarse-grained equivalent, formed from plutonic basic magma, is known as **gabbro**. Basalt or dolerite that has basic plagioclase and interstitial glass is known as **tholeiite**. Basalt often contains cavities or **amygdales** partially or wholly filled with minerals (e.g. zeolite). **Crinanite** is virtually a type of **teschenite**, an alkali gabbro.

Norite and **anorthosite** are types of gabbro, the latter being distinguished by being more than 90% feldspar, giving it a white appearance.

Acid magmas tend to be more viscous, so that they do not flow freely, and granite tends to occur in masses rather than as dykes or sills. The volcanic type of granite (**rhyolite**) occurs on St Kilda[28], but almost all our granites are plutonic and are therefore coarse-grained. Granites with very large crystals are **pegmatites**. Granite with less quartz and more feldspar is known as **granodiorite**. The names of acid igneous rocks have generally been applied rather loosely, and many 'granites' are in fact granodiorites, especially those rich in biotite and hornblende. **Trondhjemite** is a sodium-feldspar-quartz granite/granodiorite also low in ferromagnesian minerals (see Box 2.4).

If the rock has less quartz than a granodiorite, it becomes intermediate rather than acid, and is then known as **diorite**. **Tonalite** is a quartz diorite.

Where the name of an igneous rock is prefixed by '**meta-**', it means that the rock is the metamorphosed product of an igneous rock, though the prefix is often omitted for convenience. Thus **meta-anorthosite** and **metagabbro** are all **orthogneisses**.

Fragments of the pre-existing rock may be incorporated in an igneous rock as **xenoliths**, often in a partially metamorphosed form.

Migmatites are described in Box 2.3.

was one who once gave me his bag to carry to the inn by a short cut across the hills, while he walked by another road. I was wondering myself why it was so dreadfully heavy, and when I got out of his sight I was determined to see what was in it. I opened it and what do you think it was? It was stones!" "Stones!" exclaimed his companion, opening his eyes, "Stones! Well, well, that beats all I ever knew or heard of them! And did you carry it?" "Carry it! Do you think I was as mad as himself? No, I emptied them all out, but I filled the bag from a cairn near the house."[26].

A geologist of more recent vintage who was studying the Outer Hebrides Thrust pointed out in an article in the widely read scientific magazine *New Scientist* in 1970 that working in the Hebrides was fraught with "myriads of malevolent midges, strafing attacks by indignant sea-birds, the attentions of randy sheep and the effects of too many 'wee halves'"[22].

THE ARCHAEAN: THE EARLIEST ROCKS

The story of the Outer Hebrides begins in the Archaean, the earlier part of the Precambrian, prior to 2500 Ma. Not only was the Earth's interior some 2-3 times hotter than at present, but the crust was much thinner (see Box 2.5).

Today, crust consists mainly of potassium-rich granites, but during the Archaean, it was a mixture of basalt and sodium-rich granites. Continental (tectonic) plates of the Archaean survived only a short time as crust before being subducted into the mantle, so that they were still hot as they were drawn downwards, and thus melted at shallower depths than they would today[74].

Some Archaean rocks survive in the north-west of Scotland and these, like those of western Greenland, have played an important role in the deciphering of the history of crustal development.

The earliest gneisses are named Scourian, from the village in West Sutherland. Most of the Outer Hebridean Scourian rocks belong to an earlier part of the Scourian, the 'Badcallian' (Badcall is a township near Scourie), which lasted from about 2900 to 2700 Ma. Prior to the Badcallian, there must have been sedimentary and igneous rocks, as the Badcallian metamorphism acted on these to form the earliest gneisses. A later Scourian metamorphism, the Inverian (about 2600 Ma) had very little effect on the Outer Hebrides, the only unambiguous evidence of its activity being a very limited exposure on the island of Fudaidh in the Sound of Barra[4,21]. We know little of the pre-Badcallian rocks, however, and this meagre knowledge has been gained from working backwards from the Scourian gneisses. Scourian gneisses are generally thought to have been formed by the intrusion of large volumes of tonalitic and granodiorite magmas (see Box 2.2) into the lower part of a crust of pre-existing sedimentary and igneous rocks, at a depth of some 30–40km[77] (but see below). This happened under conditions of high temperature and pressure, so that there was *high-grade*

BOX 2.3 – METAMORPHISM

As the rocks of the Ness area and parts of South Harris were formed by the metamorphism of pre-existing sedimentary rocks deep in the Earth's crust, and sedimentary rocks can be laid down only on the crust's surface, it follows that there must have been a phase before the Scourian (i.e. more than 2900 million years ago) during which ancient surface rocks were drawn downwards by subduction (see Box 2.5), a process which is still going on today in some parts of the world. Deep in the Earth's crust, ancient sedimentary and igneous rocks were subjected to heat and pressure so intense that they were not only deformed physically, but many of their minerals were chemically altered – hence the name 'metamorphic' for the rocks so formed.

Pre-existing sediments subjected to intense heat and/or pressure deep in the crust may be altered to **metasediments**, also known by the old term **paragneiss**. Pre-existing igneous rocks altered to gneisses become **orthogneisses**.

Where the identity of the pre-existing rock is known, its metamorphic product may be referred to by the name of the original rock prefixed by '**meta-**', e.g. anorthosite is metamorphosed to **meta-anorthosite**. The term **metabasic** describes metamorphosed basic igneous rock.

Mild, or 'low-grade' metamorphism of sedimentary rocks produces sheet-shaped crystals, giving rocks such as **slate** or **pelite** from shale. Medium-grade metamorphism, with more heat and/or pressure (usually deeper in the crust) alters the shale to a **schist** (which retains some of the flat cleavage seen in slate). High-grade metamorphism produces a more irregularly layered **gneiss**.

Quartzite is metamorphosed sandstone, while **marble** is metamorphosed limestone.

Where metamorphism is of a very high grade, the rock begins to melt differentially, forming **migmatite,** a rock intermediate between igneous and metamorphic types. When the igneous granite dome of Lewis and Harris was emplaced, the heat caused differential melting of the surrounding rocks, melting some layers to form light-coloured **granite** (see Box 2.2) and metamorphosing the unmelted layers to a very dark-coloured gneiss, the combination being migmatite.

Low-grade metamorphism in the presence of water (**hydrothermal**) is known as **saussuritisation**, and is characteristic of the area affected by the Outer Hebrides Thrust at Lingerabay, where the **plagioclase** was altered to **epidote** (see Box 2.4). The presence of fluids affects the metamorphic grade: if dry it is **granulite facies**, if wet it is **amphibolite facies**.

Pressure applied to layers may cause them first to break up, then elongate, and Scourie dykes and 'pods' of rock may then have the appearance of a string of sausages, known to geologists as **boudins** and the process as **boudinage**, from the French *boudin*, 'a sausage'.

metamorphism on a large scale, and the Scourian is usually described as the 'granulite facies metamorphism' because of the rocks formed. Having said this, the grade of metamorphism in parts of the Outer Hebrides seems generally to have been lower than that of the Scourie area of Sutherland, so that 'granulite' grade was attained only in the south and east of the Outer Hebrides, and in parts of the South Harris Igneous Complex[39]. Recent work in Gruinard[7] suggests that the Scourian metamorphism may have involved only rocks *in situ* without any additional intrusion[25]; if this is true, some of the accepted theories of the origins of the rocks of the north-west of Scotland will have to be substantially re-written. The Scourian rocks are thought to have formed in a subduction zone. Basic and ultrabasic bodies within the Scourian gneisses are thought to represent fragments of oceanic crust which were drawn downwards by subduction, where they formed sheets in the lower part of the continental crust[49].

The Scourian is known as the 'gneiss-forming metamorphism' and is largely responsible for giving the gneiss its characteristic banded appearance[77].

THE LAXFORDIAN: THE LATER METAMORPHISM

Between 2500 and 1400 Ma, in the post-Archaean part of the Precambrian, there was a second major phase of metamorphism, peaking at around 1700Ma, the 'Laxfordian', named from the Loch Laxford area of Sutherland, of which later Lewisian gneisses are characteristic. Though some foliation would have formed during the Scourian, it was the metamorphism associated with the Laxfordian orogeny (see Box 2.5) that imposed most of the foliated pattern on the gneiss, in places developed to a wonderful pattern of successive swirling folds[58]. Laxfordian metamorphism took place in the middle-upper crust. Some of the best Laxfordian folds may be seen on the coast and foreshore of north-western North Uist between Hougharry and Foshigarry; at Geodha Ceann a' Gharaidh, to the south of Scolpaig Bay (NF72157450), the north side of the geo shows a spectacular S-shaped Laxfordian fold[48].

Between the two metamorphic phases, the Scourian and the Laxfordian, around 2500 Ma, there was a period (or periods) of great igneous activity in the Outer Hebrides, resulting in the widespread intrusion of basic dykes and sills (see Box 2.2). These are similar to the 'Scourie Dykes' of the mainland, and usually consist of bands of dark brown or black tholeiitic dolerite (see Box 2.2) a few centimetres to some tens of metres thick[78,21].

The 'Scourie Dykes' of the Western Isles may have been emplaced at greater depths than those of the mainland, up to 20 km down in the crust[30], and it has been suggested that they were intruded into a rigid crust which was under tension[72], perhaps in a situation not unlike that which gave rise to later Tertiary igneous activity (see below).

These dykes and sills serve as useful markers. Where they are relatively unaltered, the rocks could not have been much affected by the later (Laxfordian) metamorphism. Where they have been altered, the degree of deformation gives some indication of the intensity of the Laxfordian metamorphism in the area[21]. Interestingly, the Scourie dykes were first used as indicators of the stage and degree of metamorphism by Sutton and Watson in 1951[73], who also coined the terms 'Scourian' and 'Laxfordian', though it was not until 1961, with the advent of radiometric dating, that their chronology could be confirmed, as indeed it was[56]. Examples of altered and unaltered Scourie dykes from Howmore Quarry (before it was covered with boulders) are shown in Fig. 2.2. The coast of western Benbecula also demonstrates varying levels of Laxfordian folding of the Scourie Dykes, with some readily discernible S- and Z-shaped folds[49].

The Laxfordian metamorphisms covered the span between 2200 and 1400 Ma, with events possibly in the order below[56,27]:

1) Emplacement of later Scourie dykes

2) Formation of South Harris Igneous Complex

3) Early Laxfordian deformation and high grade metamorphism (1900 Ma?)

4) Early Laxfordian migmatites and emplacement of granites and muscovite pegmatites

5) Late Laxfordian 'amphibolite facies' and 'retrograde' metamorphism

6) Late or post-Laxfordian brittle folds and crush belts

In contrast to the mainland, most of the gneisses in the Outer Hebrides *were* strongly altered by the Laxfordian metamorphism: only small areas of Scourian-type gneisses, with their relatively unaltered 'Scourie' intrusions, survive[65]. Examples may be seen at Udal, North Uist (NF815780); Garry-a-Siar, Benbecula (NF756535), and Balivanich (NF761554). Perhaps the most interesting example of relatively unaltered Scourian gneisses and Scourie dyke intrusions may be seen at Ardivachar Point, South Uist (NF741463), where the gneisses were believed to be the oldest rocks in Britain, dating back more than 3000 Ma to pre-Scourian times[15] but recent opinion is that they are of late Scourian age[21]. A rock from this locality was used as the base for Parliament's gift of the Magna Carta to the US Congress on the USA Bicentenary in 1976[4]. The Ardivachar rocks are of great geological importance and should not be hammered without consulting Scottish Natural Heritage.

The above sets the Archaean and Proterozoic 'scenes' and should be recalled in dealing with the different types of gneiss described below.

GREY GNEISSES

The grey or 'undifferentiated' gneisses marked on the geological maps seem to form the main rock of the Outer Hebrides; the term describes gneisses that have not been classed as either metasedimentary gneisses or orthogneisses. The grey gneisses consist mainly of quartz, plagioclase feldspar, and hornblende (Fig. 2.3), with small quantities of biotite. Hornblende is often the main mineral of the dark bands; in close-up, the facets of the black crystals glisten in fresh exposures. Compared with the Lewisian gneisses of the mainland, fewer of those of the Outer Hebrides are basic, while more of the island rocks are granitic. Some 15% of the area of gneiss is veined by granite and pegmatite[77]. The characteristic rusty colouring of weathered gneiss is due to the weathering of the pyrite content (iron disulphide, more commonly known as iron pyrites or fool's gold)[78]. The authors of the Geological Memoir[21] have concluded that these grey gneisses represent a complex sequence of tonalite, granodiorite and granite intrusions, collectively known as the tonalite-trondhjemite-granodiorite (TTG) suite (see Boxes 2.2 and 2.5) The metabasics are the oldest rocks in the area, probably rather older than 2900 Ma, but it is believed that the intensity of the Scourian metamorphism 're-set' the radio-isotope clock[25].

Stewart Angus, July 1979

Figure 2.3. Hornblende in gneiss, Shawbost shore, Lewis. The hornblende is the dark mineral

In some areas there are large outcrops of gneisses that are obviously different from the grey gneisses. The main outcrops of these 'differentiated' gneisses are at Ness, in South Harris, and in eastern South Uist, with smaller outcrops elsewhere: these are described below.

GNEISSES OF SEDIMENTARY ORIGIN

Outer Hebridean metasediments originated from sediments deposited on the crust in pre-Scourian times. After being drawn deep down into the crust by subduction they were altered to metasediments during the Scourian. They were also subjected to later Laxfordian metamorphism, but were relatively unaffected.

In the Ness area of Lewis the grey gneisses are interleaved with layers of metasediments, including hornblende schists and semi-pelitic garnet-gneisses (see Box 2.3). In places these metasediments occur with meta-anorthosite, e.g. north of the pier at Port Skigersta (NB550622)[76]. The wonderfully photogenic 'flaggy' folding of some of the rocks at the Butt of Lewis is due more to low-temperature/high stress Laxfordian deformation than to the metasedimentary nature of the gneiss[4].

There are also small areas of metasediments in grey gneisses on the coast directly west of Doune Carloway, Lewis (NB178410)[4]. Deformation in this area and in much of Great Bernera was weak during the Laxfordian, so that many Scourian features survive[13].

The Langavat Metasedimentary Belt runs from Borve to Finsbay in South Harris, and can be followed quite easily on the ground (Fig. 2.4). Most of the belt is formed of pelites (see Box 2.3), while Chaipaval and Toe Head are rich in both biotite and muscovite micas[14,4]. The Ardvey and Quidnish peninsulas have amphibolites showing extensive alteration to epidote, actinolite and chlorite[21]. In the Leverburgh belt, one of the highlights is the presence of calcsilicate and marble bands, one of which runs NW from St Clement's Church at Rodel through Glen Rodel towards Leverburgh. There is also a thin band of marble on the NE side of Sta Bay west of Borve, on the west coast of South Harris NG029949[21]. The 'crystalline limestone' of Rodel is said to be characterised by the presence of white nodules of diopside[57].

Metasediments also occur on Ensay, Killegray and Pabbay in the Sound of Harris, and in the Uists (Fig. 2.1). The best example in this area is probably at Claddach Kirkibost, North Uist (NF786650) where garnet-rich metasediments crop out on the beach[4]. Garnet-rich metasediments may also be seen on the foreshore just north of Rubh' Aird-Mhicheil, South Uist (NF730334)[4].

CLEITICHEAN BEAG GNEISSES: THE PENTLAND ROAD KNOLLS

The zone of low Laxfordian deformation around Carloway also contains unusual Scourian igneous intrusions, virtually unaltered by the Laxfordian metamorphism. These 'Cleitichean Beag' gneisses crop out as two peaks at NB268372 and NB272368 on the south-west side of Loch nan Cleitichean, near the fork in the 'Pentland Road'.

If you go to the fork near the west end of the Pentland Road and walk either northwards or north-eastwards for a few hundred metres, you will come upon a couple of rocky knolls. Closer inspection of the outcrops reveals a rock surface which closely resembles an egg-box in appearance. There are smaller outcrops of Cleitichean Beag type gneisses elsewhere in western Lewis and north-west Harris, notably on Great Bernera, though most of these have been affected by later Laxfordian events. The outcrops at Cleitichean Beag have been dated at around 2440±60 Ma, i.e. late Scourian. These ultrabasic rocks (see Box 2.2) have an unusual mineral composition containing pyroxenes, olivine, and plagioclase feldspar, with minor amounts of hornblende and biotite[41,4]. They resemble coarser-grained metadolerites, to the extent that in parts of Pairc and North Uist they may be difficult even for geologists to distinguish[21].

OTHER GNEISSES OF IGNEOUS ORIGIN: THE FORMATION OF SOUTH HARRIS

The rocks of the South Harris Igneous Complex are believed to have been intruded into the metasediments as a sheet-like mass and subsequently meta-morphosed, possibly around the time of the intrusion of the Scourie Dykes[14] or, more probably, in the early Laxfordian, around 2250 Ma[56,39]. The Complex may have developed at a great depth in the crust[78]. The Complex represents one of the few examples of Scottish Laxfordian igneous activity, though there are meta-igneous granite sheets around Loch Laxford in West Sutherland, and there was widespread contemporary igneous activity in Scandinavia, Greenland and North America[49], which at that time were located near the Outer Hebrides, before they were separated by continental drift (see Box 2.5). Six rock types, dated to about 2200 Ma[10] have been identified in the Complex, all of them intermediate, basic or ultrabasic in character, in contrast to the acid rocks which make up the greater part of the Lewisian outcrop. They are coarse-grained, highly resistant, and mostly very dark in colour, forming much of Chaipaval, Bleabhal and Roineabhal[65] (Fig. 2.4). Roineabhal consists mainly of white meta-anorthosite which is believed to have been formed by high-pressure metamor-phism and is strongly folded[78]. The meta-anorthosite contains bands of horn-blende and pyroxene, xenoliths of metasediments, and numerous basic (Scourie-

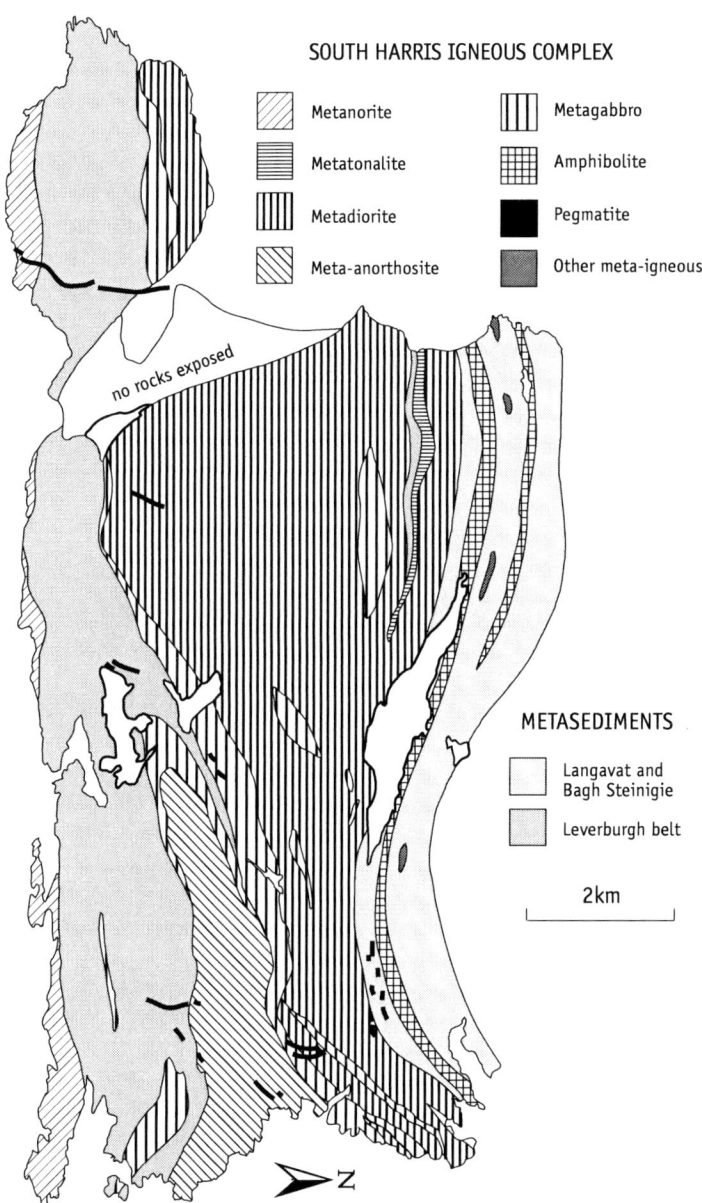

Figure 2.4. Simplified geology of South Harris (after British Geological Survey 1982. Geology of the South-west Part of Harris. *British Geological Survey, Keyworth). Reproduced by permission of the British Geological Survey.*

type) dykes, the last being plainly visible from the Finsbay-Rodel road[14]. With the exception of the meta-anorthosite, the rocks of the South Harris Igneous Complex are difficult for the amateur to distinguish in the field.

The South Harris anorthosite covers an area of some 7–8km^2, and is the largest expanse of this rock in the UK, though there are areas of 1800km^2 in Norway and over 3100km^2 in the Adirondacks in the USA[46]. In broad terms, the Harris anorthosite can be divided into two types, with the boundary between the two located roughly along the line of the Rodel-Finsbay road. To the west, forming most of the bulk of Roineabhal, is the 'pink anorthosite', where the colouration is due to the presence of haematite within the feldspar. Between the road and the sea lies the 'white anorthosite', where the plagioclase feldspar has been replaced by minerals such as zoisite, a form of epidote (see Box 2.4). The white anorthosite should not be confused with the intense white colour of weathered surface layers: this is thought to be due to leaching which is a recent process involving water percolation. The alteration of the pink anorthosite to the white form is known as saussuritisation, and is thought to be associated with movement in the Thrust Zone (see below).

Most of the South Harris Igneous Complex consists of metadiorite (or tonalite), a very dark coloured rock, while the southernmost part of the Complex, comprising much of the Harris coastline of the Sound of Harris, the northern part of Ensay, and probably much of the rock underlying the Sound of Harris, consists of meta-norite (see Box 2.2) which is allied to meta-anorthosite, being a product of high-pressure metamorphism, but is much darker in colour[4,78]. The best outcrops of the norite are thought to be those at Renish Point NG042826, on Chaipaval, and at Strond NG026827. Garnets are abundant in the section between Strond and Renish Point. The order of intrusion is thought to have been: gabbro, anorthosite, norite, diorite and tonalite[21].

An ultrabasic intrusion at Scara Ruadh (NG057884) (Fig. 2.5) near the southern tip of Loch Langavat in South Harris is most unusual for a meta-igneous rock in that it displays a layered structure[4]. This layering consists of serpentinite bands 10–250cm thick and chlorite/tremolite layers up to 60cm thick[21]. A smaller ultrabasic intrusion on the south-western shores of nearby Loch Meurach (NG060877) consists of serpentine with veins of talc and 'asbestos'; chlorite schists and chlorite-actinolite schists are also present[14,21]. Some 650m north of Manish on the island of Scarp, in the hollow of the south slopes of Sron Romul, there is a conspicuous outcrop of foliated gneiss which consists of "a felted mass of amphibole in which are embedded blades of green actinolite"; clusters of these blades may be 5cm in diameter and have a fibrous appearance[38]. The asbestos rock is referred to as *Clach a' Chomharra* by Angus

Stewart Angus, November 1996

Figure 2.5. Scara Ruadh, South Harris

Duncan in his *Hebridean Island: memories of Scarp*[17], and a colour picture in this book shows the rock topped by a glacial erratic of granite. A sill-like mass at Eilean Glas, Scalpay, has been described as containing veins of steatite (the rock formed from talc, often known as soapstone) and fibrous chrysotile[37]. The Scara Ruadh rock is part of a ridge of serpentine which extends the entire width of South Harris, of which the Victorian geologist Professor Heddle said there was enough asbestos at the east end to "load an Indiaman"[32].

The boundaries between the metasediments and the grey gneisses and the Igneous Complex are often abrupt: possibly the separate origins of the three types persisted through the Laxfordian metamorphism[76].

The Corodale gneisses of eastern South Uist consist mainly of garnet, plagioclase, diopside and brown hornblende (see Box 4)[4]. These basic gneisses are more resistant to erosion than the grey gneisses, and form the hills of Triuirebheinn, Stulabhal, Beinn Mhor and Hecla[21]. They are believed to be older than the rocks of the South Harris Igneous Complex, having formed some 2700-2900 Ma ago[80]. To the non-geologist, the rock itself seem unremarkable, but the legacy of thrusting, which brought the Corodale gneisses into contact with the 'grey' gneisses, is very striking, with a contact zone of pseudotachylite or phyllonite (see Earth Movements, below).

Neither the South Harris Igneous Complex nor the Corodale gneisses have any equivalent on the Scottish mainland[56].

BOX 2.4 – MINERALS

All rocks are made of minerals, and the classification of a rock depends on its mineral composition, which in turn is largely determined by its mode of formation. Our rocks are composed almost entirely of **silicates**, though some contain significant amounts of other minerals, such as **calcite**, a form of calcium carbonate found in the conglomerate of the Stornoway Formation, and **magnetite**, an iron compound occasionally seen in granite, which will deflect a compass needle held at close quarters. **Haematite** is ferric oxide, an iron ore.

Silicates

Classification of silicates relies on a knowledge of the molecular structure, and is based on the number of shared oxygen atoms per silica tetrahedron. Anyone requiring more detail than that given in the summary below should refer to a specialised text book.

Quartz is one of our commonest minerals, sometimes occurring as very large crystals in granite pegmatite (see Box 2.3), and is readily identified by its milky white colour and by its hardness; unlike calcite it cannot be scratched with a pin.

Feldspars are also very common, and are very variable. Like quartz, they are not base-rich, and do not weather readily. The pink colour of granite is due to an abundance of **orthoclase** feldspar. **Plagioclase** feldspar is opaque white or grey, and occurs in a great variety of forms, e.g. **andesine** and **labradorite**, the latter being the main feldspar of the pink anorthosite of the South Harris Igneous Complex. The varieties of plagioclase cannot be identified in the field, which is unfortunate, as they are crucial to the classification of many igneous rocks.

Pyroxenes include **diopside** and **augite**, and are found most commonly in basic and ultrabasic rocks but also occur in high-grade metamorphic rocks. **Olivines** are closely related to pyroxenes and are green or greenish brown in colour, being frequent in basic and ultrabasic rocks. Olivine-rich rocks are known as **peridotites**. It is pyroxene and olivine which contribute most to the very dark colours of most basic and ultrabasic rocks.

The commonest of the **amphiboles** is **hornblende**, a greenish-black mineral

---

GRANITE-MIGMATITE COMPLEX: THE HILLS OF HARRIS AND UIG

During the late Laxfordian, about 1700 Ma ago a huge dome of granitic magma forced its way into the gneisses, forming what is now Uig, central North Harris, and northern South Harris, covering an area of some 420km^2 (Fig. 2.1)[21]. Granite is irregularly distributed within the granite-migmatite complex[53]. Gneisses adjacent to the granite were migmatised (see Box 2.3) and have the appearance of grey gneiss veined with granite[65]. Granite is composed mainly of

frequent in many igneous and metamorphic rocks, including many gneisses and schists. Hornblende schist is also known as **amphibolite**. The 'asbestos' of Scarp is **actinolite**, an amphibole. The amphiboles are base-rich and weather readily. **Tremolite** is an amphibole, often known as **tremolite-actinolite**.

Micas are readily identifiable by their flaky texture. While they occur as flecks in schists and gneisses, they are most spectacularly seen as huge crystals, several centimetres across, in pegmatites. **Biotite** mica is black or very brownish black, while **muscovite** is clear. The biotites of pelitic and semi-pelitic metasediments tend to have a purplish tint whereas those of other gneisses are the normal brownish-black[12].

Garnets occur as dark pink crystals in gneisses and schists and, to a lesser extent, in granites. Our garnets are not of gem quality.

Epidotes are greenish minerals occurring in igneous and low-grade meta-morphic rocks though one form, **zoisite**, is white or light grey, and contributes to the white colour of the anorthosite of South Harris (which is >95% plagioclase).

Serpentine is formed by the metamorphosis of olivine or pyroxene in the presence of water, giving a light green mineral, common in the basic and ultrabasic rocks of the South Harris Igneous Complex. The 'asbestos' of the Igneous Complex is a form of serpentine more accurately known as **chrysotile.**

The metamorphosis in the presence of water of the **ferromagnesian** or **mafic** minerals (containing iron and magnesium) such as olivine, augite, hornblende and biotite, may form **chlorite**, which is soft, green and flaky, but the flakes are not so flexible as those of mica. Chlorite may contain high proportions of the extremely soft mineral **talc**, though it is green rather than the familiar white form found on the shelves of the pharmacy. **Lamprophyre** is a medium-grained rock rich in ferromagnesian minerals, often containing large crystals of these minerals.

Zeolites are crystals often found in amygdales.

Some of the rare minerals mentioned in the sections on Rockall and the Loch Roag monchiquite are beyond the scope of a popular text: those wishing to know more should consult a good geological dictionary.

---

quartz, mica and feldspar, the type of feldspar determining whether the overall colour of the granite will be pink, white or grey. The grey gneisses within the affected zone tend to be exceptionally coarse-grained[78]. One outcrop of pegmatite at Garry-a-siar, Benbecula (NF756534) contains feldspar crystals up to 2m long[21]. Xenoliths of grey gneiss occur within the granite-migmatite, one of the best examples being at Loch Raonasgail in Uig (NB033274) where a large, angular xenolith of grey gneiss may be seen in the coarse-grained granite[53]. Exposures in road cuttings show increasing degrees of migmatisation west from Grimersta[21].

Pegmatite veins were injected at about the same time as the granite-migmatite complex was formed, and pegmatite dykes are frequent on the west coast of the islands$_{78}$. One of the best examples of a pegmatite dyke is the conspicuous vein on the east flank of Chaipaval$_4$ which can be seen from a great distance. It has been said that the Chaipaval pegmatite vein is the longest in Scotland$_{32}$(Fig. 2.6): it consists of quartz, plagioclase, microcline and muscovite (white mica) with minor biotite (blackish mica) and magnetite. The 'books' of muscovite formed by the flaky crystals are locally spectacular. The dark pink potassium feldspar is easily distinguished from the adjacent pale pink plagioclase; locally, there is pale green epidote. Some of the minor inclusions here are mildly radioactive, and the spoil heaps have been well picked over by people hunting for gemstones such as beryl and tourmaline, but in two hours a professional geologist and I were unable to find anything remotely resembling a gemstone.

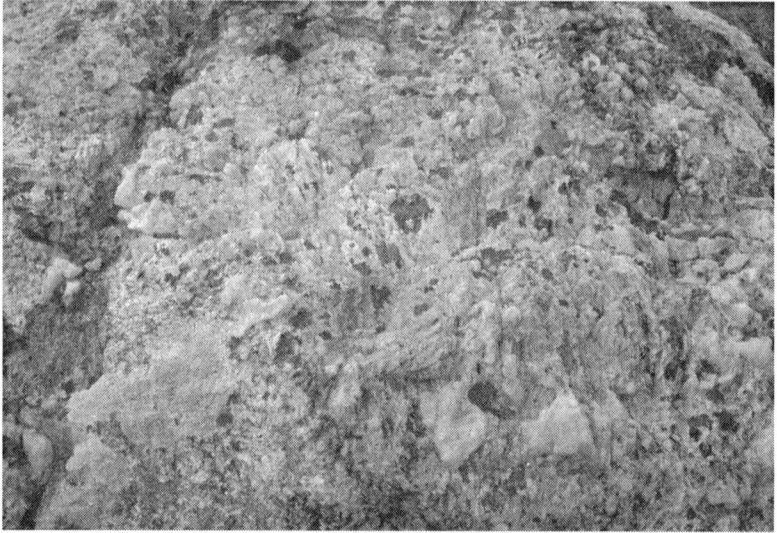

Stewart Angus, November 1996

Figure 2.6. Chaipaval Pegmatite

The quartz in the pegmatites is always white or transparent, except in the centre of western Scarp, where the quartz is a deep purple colour$_{53}$; on Sletteval, where quartz in the core of the pegmatite may have a bluish purple tint$_{54}$, and on the summit of 'Bosnival', assumed to be Bhoiseabhal NG0487$_{46}$ where the pegmatites contain smoky quartz. Magnetite may occur in these pegmatites in

Stewart Angus, June 1988

Figure 2.7. Pegmatite-gneiss junction on Sletteval

masses up to 15cm across[21] (and can be readily identified by its deflection of a compass needle at close quarters) and it has been suggested that in some areas magnetite may make up as much as 50% of the rock over an area several metres in diameter[53]. The pegmatites may contain small quantities of radioactive minerals[21].

Where three granite veins converge on the island of Stromay in the Sound of Harris, there are said to occur "huge crystals of nacreous [mother-of-pearl] feldspar"[32].

The Laxfordian granite-migmatite complex forms the imposing cliffs of Mangersta, Uig, while inland it forms the bare terrains of central South Harris and hills such as Suainaval in Uig[65].

EARTH MOVEMENTS

The Scourian events took place deep in the crust, the Laxfordian at middle and upper levels, and erosion of the Laxfordian rocks is known to have contributed to the formation of Torridonian sediments on what is now the NW mainland of Scotland, and so these Laxfordian rocks must have been exposed by 1,100 Ma[77]. There has thus been considerable uplift over the Precambrian, associated with phased activity around the margins of tectonic plates (see Box 2.5).

BOX 2.5 – PLATE TECTONICS

Plate tectonics – widely if inaccurately referred to as 'continental drift'- have been responsible for much of the geological activity which led to the formation of what is now the Outer Hebrides.

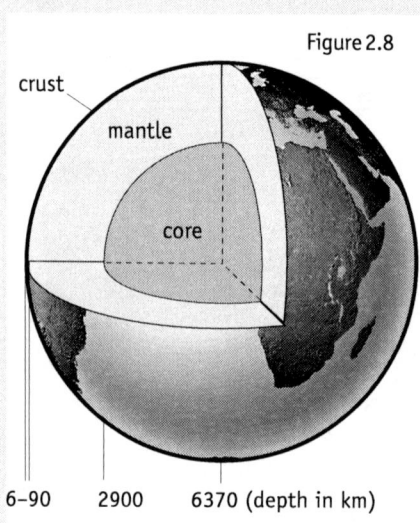

Figure 2.8

6–90   2900   6370 (depth in km)

To understand the mechanisms of plate tectonics, it is necessary to envisage a cross section of the earth, from the centre to the surface. This radius measures 6370km, and is made up of three layers: the innermost **core**, a middle **mantle** and an outer **crust** (Fig. 2.8). The crust is seldom more than 60km thick, and 'floats' on the semi-molten mantle as a number of rigid tectonic plates. [The crust floating on the mantle is the mechanism responsible for the vertical land depression and recovery referred to in conjunction with glaciations in the next chapter]. Apart from the deposition of sediments, all geological activity takes place around the margins of these plates.

The boundary between the crust and the mantle is named the Mohorovičić discontinuity, after its discoverer, but it is usually known as the **Moho**, as even geologists find his name a mouthful.

Because of irregular convection currents in the mantle, the tectonic plates move relative to each other. Where the plates move apart, they usually do so from a mid-Ocean ridge, where mantle magma is extruded and solidifies to form new oceanic crust (Fig. 2.9), so that the plate boundary is **constructive**, as in the mid-Atlantic, where Europe and the Americas are moving apart at a rate of about 1cm a year. Oce-

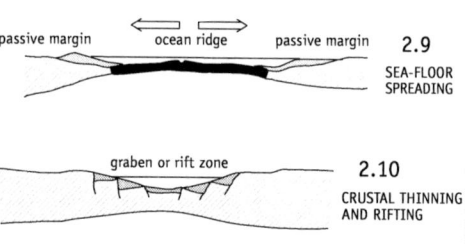

anic crust is denser than continental crust and is usually only some 6km thick. Oceanic crust forms the floors of the deep ocean, and should not be confused with the shallow continental shelf, which is merely underwater continental crust. Where a tectonic plate parts slightly but does not actually split sufficiently to form new oceanic crust, a **graben** or **rift** zone develops (Fig2.10). It has been suggested that the formation of the Minch graben was a first attempt at the opening of the Atlantic.

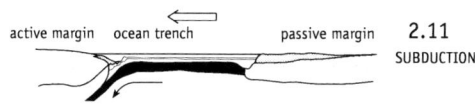

active margin   ocean trench   passive margin   2.11
SUBDUCTION

If some plates move apart, it follows that somewhere this is matched by plates moving together. Where plates collide, the plate boundary is said to be **destructive**; the continental crust rides over the denser oceanic crust, and material from both types of crust is drawn down into the mantle by **subduction** (Fig. 2.11).

When wet crustal material (e.g. oceanic crust and its overlying sediment) is heated at depths of 80–300km, the heat drives off water and other volatile components, which act like a flux in a foundry, melting the surrounding material at a lower temperature than usual, giving rise to the sort of explosive eruptions still seen at plate boundaries today. The interior of the Earth generates its own heat, and magma may accumulate above such a heat plume, ultimately forming domes of granite. During the late Archaean, subduction of wet basalt oceanic crust produced tonalitic magma on a large scale, which then rose upwards, while thermal mantle plumes added to the continental crust from below, so that together these two mechanisms gave rise to a rapid build-up in the thickness of continental crust. When you realise that most of the Earth's surface is not only covered by water, but most of the oceanic crust is submerged to a depth of 4km by sea water, the build-up of continental crust during the late Archaean was instrumental in the formation of the continental plates we see today.

Today, the crust of the Earth consists largely of potassium-rich granite, but until about 2,500 Ma it was a mixture of basalt and sodium-rich granites, making up the tonalite-trondhjemite-granidiorite (TTG) suite (see Box 2.2).

Measurement of minerals and radio-isotopes can reveal the pressures and temperatures (and thus depths) at which rocks were formed, at a time when the Earth was cooling rapidly. The Late Archaean continental collisions contributed to mountain-building or **orogeny** above the collision zone (as in the Andes today), and there was wide-scale **high grade** metamorphism and folding associated with the high temperatures and pressures below these collision zones. It should be noted, however, that there remains considerable controversy over the use of radio-isotopes for the interpretation of metamorphic history, as indicated by G.Rogers and R.J.Pankhurst[63] in a major review of the use of dates in the metamorphic areas of Scotland:

> Not only can the structural geologists not always agree about which structures are being dated (e.g. Inverian or Laxfordian) but many of the ages so far produced are by and large of dubious chronological validity.

By the time of the Laxfordian events, the crust had cooled, and instead of a high grade metamorphism, rocks were stretched and folded to varying degrees, but it was largely a **low-grade** or **retrograde** (lower grade than previous) metamorphism, and is largely known as the 'amphibolite facies metamorphism': in areas of low grade metamorphism pyroxenes are progressively replaced by hornblende and biotite.

The end of the Silurian period (c. 400 Ma) was the climax of the Caledonian Orogeny, and also marked the collision of Scotland and England, which had previously been separated by the Iapetus Ocean.

The Outer Hebrides Thrust Zone represents a later movement, where eastern gneisses were pushed sideways over those of the west for distances of up to 10km[78], resulting in the severe crushing or 'mashing' of the former. The line of the Thrust extends south from Tolsta Head to Sandray and beyond, hugging the east coast for its entire length south of Loch Odhairn (Fig. 2.1). Seismic work offshore suggest that the Outer Hebrides Thrust extends down through the crust to the Moho, the boundary between the Earth's crust and the mantle (see Box 2.5), which lies at a depth of about 27km in the Outer Hebrides[67]. The Thrust is believed to represent the longest surviving crustal movement in the British Isles[8]. Opinions differ as to the timing of movement(s) on this thrust, with some recent work suggesting that the fault may even predate the Laxfordian pegmatites and the Scourie dykes[40], but the latest studies confirm earlier estimates that the thrust was active from 1700 Ma onwards[8]. The heat produced by friction when the two masses of rock moved relative to each other (within the crust) was such that the rocks in the contact zone melted and were injected into the adjacent gneisses. The melted rock solidified very quickly to form pseudotachylite, a hard, black, occasionally glass-like rock with a flinty fracture which typically forms small cubes when broken[21]. It can also cause strong deflections of magnetic compasses[23], and it is interesting to note that the famous SS *Politician* which was the inspiration for the SS *Cabinet Minister* in Compton Mackenzie's *Whisky Galore* grounded on rocks of pseudotachylite in the Sound of Eriskay in 1940. Magnetic reefs have been blamed for many shipping disasters in the Outer Hebrides, but pseudotachylite is only capable of deflecting a compass needle at a distance of a few feet[22]. One of the largest masses of pseudotachylite yet identified occurs on a remote hill in the extreme south-east of South Uist, Maol na h-Ordaig (NF840148), the southern slopes of which consist almost entirely of this rock. Another major exposure occurs on the northern slopes of Beinn Ruigh Coinnich, just east of Lochboisdale[23]. Associated with the pseudotachylite is a black or green 'flinty crush' more properly known as ultracataclasite[79], which forms the rock on which Kismul [Kiessimul] Castle stands in Castlebay[81]. Where it flowed into the mashed gneiss the pseudotachylite welded the fragments together as a breccia.

Once formed, pseudotachylite is stronger than the surrounding gneiss, for all practical purposes 'welding' the thrusted rock to the underlying rock by friction, as in the Corodale gneisses. Subsequent movements take place in the adjacent gneiss, and thick layers of pseudotachylite may be built up by successive movements[21]. The very hard rock which persists as the North and South Lees and Eaval was formed in this way[65]. The best exposure of mashed gneiss welded by pseudotachylite occurs on Greian Head, Barra (NF6504)[4]. Pseudotachylite/breccia scarps also contribute to hills at Beinn Mhor and Hecla in South Uist (where the Thrust forms the western boundary between the Corodale gneisses and the grey gneisses) and on Feirihisval (NB301147) and Beinn Mholach (NB356387) in Lewis[21].

The British Geological Survey *Memoir* authors concluded that after the first phase of geological activity the thrust may have moved intermittently over a period of several hundred million years, with a second phase of thrusting and uplift around 1150 Ma[21], a final date which is backed up by the most recent estimates. The contact zone of this second phase of movement is marked by narrow bands of mylonite (= phyllonite)[65], a pale-coloured schist-like rock best seen on the relatively inaccessible east coast of South Uist, where it forms most of the rock outcrop on the island of Stulaidh[5]. Chlorite is the dominant mineral in the mylonite[64]. A search around the ruins of Ormiclate Castle in South Uist may still reveal fragments of this flaggy rock, giving substance to the tradition that the mylonite of Stulaidh was once quarried to provide roofing material[35]. In recent years Stulaidh mylonite has been used to construct at least one fireplace in Benbecula. A fine outcrop of mylonite may also be seen at Eilean Glas, near Ranish (NB422240), where a well-marked thrust plane is overlain by "platy grey, pink and striped mylonites"[57].

Pseudotachylite is absent from most of the world's fault zones[21] and the Western Isles are famed among geologists for their spectacular examples of this rock. Whether the Thrust was activated on one or more occasions, the pseudotachylite was formed during a series of earthquakes, while the mylonite is associated with slower earth movements[77].

SEDIMENTARY ROCKS

Stornoway Formation

With the exception of some small exposures on the Shiants (see below) the only sedimentary rocks unaffected by metamorphism found in the Outer Hebrides are the conglomerates and sandstones of the Stornoway Formation which occupy a total terrestrial area of around 50km^2 (Fig. 2.1)[21]. Most of the Formation consists of conglomerate – rounded pebbles in a sandstone matrix. The white mineral calcite is locally frequent in the deposits and, being soft, probably accelerates the erosion of this already friable rock.

The Second (New) Statistical Account for the Parish of Stornoway of 1833 tells of a spectacular cave in the conglomerate:

> There are two caves about the distance of an English mile from Gress House, both of which are spacious; but the Seal cave is the most remarkable in the island. It is about a furlong in length from the entrance to high water-mark in the interior. Its height and breadth are variable. The cave at the mouth, is about ten feet wide: it gradually decreases to four feet in breadth; and after this, it widens

and terminates in a spacious semicircle, irregularly arched, and containing a basin of water. Here, the roof is very lofty, and resplendent when viewed by torch-light. Beyond the margin of the basin, is a sandy and gravelly beach, very pleasant and acceptable after such a dark navigation. There is a small apartment in the interior, which by torch-light produces a fine effect; the pearly icicles of stalactite suspended from the roof, reflect the light as from so many diamonds. The sides and roof of the cave are lined with this concreted matter.

Other accounts by travellers and colleagues who had visited the cave (north-east of Gress beach at NB512418) have confirmed these observations (Fig. 2.12). The cave is quite difficult to locate, as the entrance is somewhat nondescript, and it is necessary first to navigate an outer cave which narrows to the extent that the only a canoe or a swimmer could pass. The easiest option is a dry suit, and any lighting equipment and cameras must not only be protected from the sea but also from the water which drips constantly from the roof. An added hazard is the presence of seals. There are three linked chambers in the interior, and the drops of water which accumulate at the tips of the hollow stalactites glisten in the torchlight. There are even a few stalagmites, but perhaps the most spectacular features are the sheets of calcite at the rear of the cave, where it narrows to a ledge. Calcite is frequent in the conglomerate, and is slightly soluble in water underground but becomes insoluble on contact with the air, and is then deposited, forming the stalactites, stalagmites and sheets.

David Maclennan, July 1996

Figure 2.12. Interior of the Gress Seal Cave

The beds of the Stornoway Formation overlie the Outer Hebrides Thrust but have not been affected by it, and so they must post-date the Thrust$_{68}$, thus establishing beyond doubt that they cannot be related to the Torridonian

Stewart Angus, June 1983

Figure 2.13. Junction of conglomerate with Lewisian Gneiss at Garrabost, Point. The conglomerates are on the right and centre, the Lewisian gneiss to the left.

(Precambrian) sediments of the north-western mainland of Scotland, as had been supposed in the past. The total thickness of the formation (but not any individual part of it) is nearly 4km, so that it is one of the thickest conglomerates in the British Isles[68].

The gneiss pebbles of the conglomerate seem to have originated from a mountain range to the west of what is now Broad Bay, and the beds are mainly in the form of overlapping alluvial fans associated with the rivers which flowed from the north and west, probably during the Permian and Triassic (280–200 Ma), when the land to the north and west included what are now parts of Greenland and Canada. Similar sediments lie on the bed of the North Minch[68]. No fossils have ever been found in the rocks of the Stornoway Beds, though possible root-bearing or burrowed horizons have been identified in the finer-grained deposits on the east side of Stornoway Harbour and between Melbost and Aignish[69,21].

The best exposures of the Stornoway Formation are on the coast between Aignish and Garrabost, the latter being the best place to see the junction between the conglomerates and the Lewisian basement (Fig. 2.13)[4], though a scramble down a steep slope is required to gain access to this exposure. There are also reputed to be good exposures of this junction below the Laxdale Bridge, in the bed of the Allt Roanadale (NB5143) some 300m from its mouth, and near the geo north of Sheilavig Mor (NB516434)[38].

Some of the pebbles within the conglomerate are epidotite, being pale green epidote streaked with dark red haematite. These pebbles may be strikingly beautiful, to the extent that they may be polished and mounted in jewellery as 'Lewis Pebbles'[69].

Shiants shales

The only other sedimentary rocks occurring in the Outer Hebrides are the Lower and Middle Jurassic shales of the Shiant Islands, though even these have been subjected to thermal metamorphism by later igneous activity, and occur as thin beds of baked shale interbedded with the sills (see next section). The shales are thought to have been formed from deposits of marl with some dolomite[21]. The thickest deposits are on Eilean Mhuire, where a bed of about 19m thickness crops out on the higher parts of the island. There are about 9m at sea level on the NE side of Garbh Eilean and 5m on the NW foreshore of Eilean an Tighe[21]. Some fossil belemnites and ammonites have survived this metamorphism; only a few fossils have been collected, and most are altered or otherwise difficult to identify, but the commonest species seems to be the ammonite *Dactylioceras anguiforme*[59]. Belemnites are bullet-shaped remains of the inner skeletons of relatives of squids and cuttlefish. These are the only terrestrial exposures in the

Outer Hebrides of the Jurassic rocks which form a large proportion of the bed of the Minch (see below).

If the Outer Hebrides Thrust zone really was the plane of a low-angled fault, the gneisses would have been buried by several kilometres of Mesozoic sediments which would have been eroded away during the late Jurassic and Cretaceous periods, but there is no evidence to suggest that such sediments ever existed[66].

LATER IGNEOUS ROCKS

A number of minor volcanic dykes were intruded during the Permian and Carboniferous periods (from 330 to 230 Ma) before the Stornoway beds were deposited. The majority of these dykes are only 0.5–1.5m wide (exceptionally 4m) and they consist of camptonite (82%), basaltic camptonite (9%) and monchiquite, all forms of lamprophyre[62] (see Box 2.4). About 50 of these have been mapped in the Outer Hebrides, mainly in northern Barra and on the east coast of South Uist[5]. Most of the quartz-dolerite dykes, of which there are at least 10 in northern Barra and in the extreme south of South Uist, are also of Permo-Carboniferous age; these weather to a distinctive brown colour and are 3–45m thick[21]. A monchiquite dyke in the Loch Roag area contains a range of xenoliths and one of the richest mineral assemblages of any camptonite or monchiquite dyke in Scotland, including biotite, apatite, wehrlite, mica and clinopyroxene. The most frequent mica here is phlogopite, which is dark brown, and usually occurs in circular masses. Semi-precious stones include corundum and zircon. This site featured on the front page of the Scottish national press in 1995 when it yielded the largest sapphire (a type of corundum) ever found in Scotland. A previous exercise by academics involving the week-long excavation of a road using heavy equipment provided only three stones, illustrating the rarity of the gems. Any stones, of course, remain the property of the landowner.

When the sediments of the Stornoway Formation were being laid down, Britain was situated south of the equator, in the middle of a vast super-continent. This super-continent split, with fragments moving gradually apart by continental drift to become the great continental land masses we know today (see Box 2.5).

Until 60 Ma ago Greenland was near the Outer Hebrides. As the North Atlantic opened, Greenland and North-east Canada moved north-westwards and westwards[65]. This movement was marked by considerable igneous activity not only on the diverging continental margins but also well within the boundaries of the continental crust around the British Isles[19].

Most of the igneous activity in north-western Scotland is believed to have occurred between 63 and 52 Ma ago, during the Lower Tertiary period[19].

St Kilda and Rockall are remnants of large Tertiary igneous complexes. The islet of Rockall consists of granite, most of which is alkaline, with 53% feldspar, 23% ferromagnesian minerals, 22% quartz and 2% accessory minerals (by volume). Some veins within the granite have a ferromagnesian mineral content of 68% and have been termed 'Rockallite'. Cavities within the rocks are lined with rare minerals with splendid names like bazerite and monazite[20]. So much of Rockall has been removed by visitors (two donated chunks are displayed in my own office) and the need for a level platform for the navigational light that the latter syllable of the name may no longer be entirely accurate! Rockall possibly formed from the partial melting of the upper mantle during the cooling stages of widespread igneous activity about 52 Ma ago[20]. That this tiny islet, with an area of only 624m^2 protrudes above the waves at high water, and is thus by international definition an island rather than a rock, is of enormous legal significance, in that it gives the United Kingdom ownership of the surrounding territorial waters. Rockall is part of Harris by virtue of the Island of Rockall Act 1972[6], though it was formally annexed by the Crown in September 1955, in the fear that a foreign power would establish a tracking station for then proposed missile range in South Uist[60]. A local claim predates this, however: according to the Coddy, Ruari an Tartair searched for 'Rocabarraidh' at the beginning of the 17th century, in the belief that its inhabitants owed him allegiance[44]. The geological interest rears its head again, in that it is the sole reason for the notification of Rockall as a Site of Special Scientific Interest in 1975, and again in 1984, without protest from other nations. Neither Ireland nor Denmark seem to dispute the UK's ownership of Rockall *per se* but have reserved their position with regard to the offshore resources. An article in the august legal journal *The Scots Law Times* published in 1985 could not resist a reference to a private Irish expedition of 1983, which set off to claim Rockall for Ireland, but returned in embarrassment, having failed to find the rock![60]. The same article goes into some detail regarding the obscure legal background to legislation affecting Rockall, which is apparently of importance if no one disputes the right of the UK to carry out administration affecting the territory. It seems that a Scottish solicitor, Daniel C. Gardner, realised no one had claimed title to the island and attempted to do so himself, but was apparently thwarted by the Crown. The Law of the Sea Convention refers to "rocks which cannot sustain human habitation or economic life of their own", which would affect some legal claims. The 'human life' angle was explored by the adventurer Tom Maclean, when he took up residence in a tiny plywood cabin on Rockall for forty days in May 1985[60], while the presence of a light installed by the Northern Lighthouse Board in 1971 is arguably 'economic activity'[6]. Tom Maclean applied for and received planning permission to set up his hut, despite an objection from a Mr Sean D. Dublin Bay-Rockall Loftus, who claimed to be the Chairman of the Island of Rockall Trust and legal adviser to the Dublin Bay

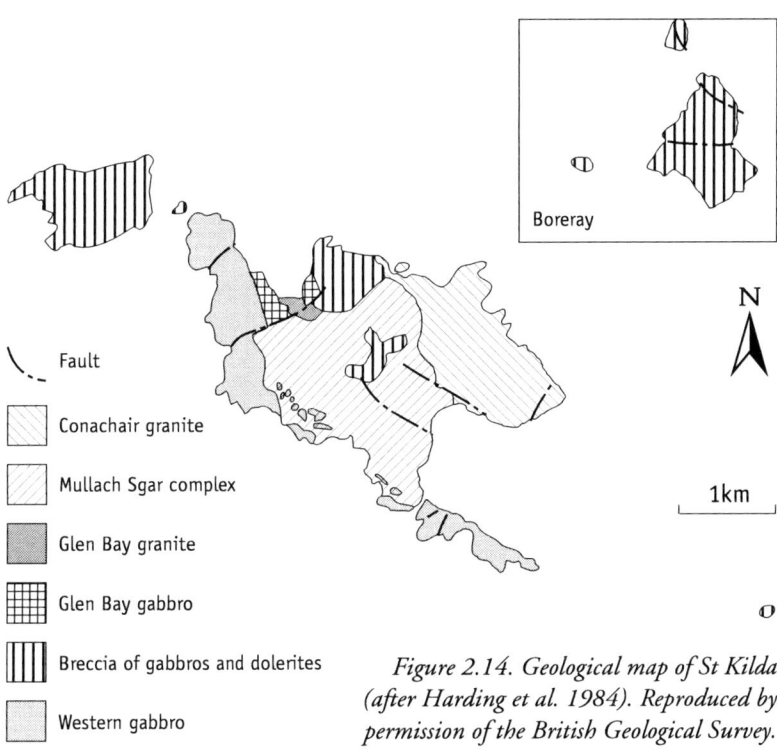

Figure 2.14. Geological map of St Kilda (after Harding et al. 1984). Reproduced by permission of the British Geological Survey.

Democratic Society. Two individuals made inquiries to the Western Isles Licensing Board in respect of their intention to set up an optic to serve drinks to Tom Maclean[60], and though no application was made, one of the Board's advisers was tempted to recommend refusal on the grounds that the application site did not match the usual car parking requirements for licensed premises.

St Kilda displays a complex range of Tertiary rocks within its 850ha (Fig. 2.14), representing various stages of intrusion associated with a volcano, dating from between 63 and 55 Ma, when northern Europe was located some 19° south of its present position. The volcano is thought to have exploited a plane of weakness in the Earth's crust.

The oldest rocks now found on St Kilda are the Western Gabbros which are layered, possibly as a result of sorting during magma flow or solidification. The weathered surfaces of the gabbros on the higher parts of Ruaival appear speckled because of small inclusions of white plagioclase crystals within large grains of black pyroxene.

Soay, Boreray, Levenish and Glacan Mor on Hirta consist of a breccia of gabbros and dolerites, possibly representing the transport and mixing of large solid blocks of basic rocks by a basaltic magma, probably at the base of a volcano. In places, amygdales are filled with green amphibole, chlorite and clay minerals.

The next rock to be formed was the Glen Bay Gabbro, which crops out on the sides of Glen Bay, where it is split by the later intrusion of the Glen Bay Granite. The chief feature of interest here is the very unusual thick glassy margin caused by quenching, possibly as a result of a large block of cold breccia collapsing into the magma chamber near the surface of the crust. The effects of chilling are believed to extend 100–120m into the gabbro.

The Mullach Sgar Complex makes up most of the southern part of Hirta and the NW of Dun, and represents four phases of intrusion, the mixed acid and basic rocks associated with these giving rise to very complex geology on the west of Ruaival and on the cliffs of Caolas an Dùin.

The youngest of the main intrusions, dated at 56 Ma, is the Conachair Granite, which makes up the east of Hirta, including Oiseval. This is usually the first rock encountered by visitors, as it crops out around the pier, and was used to build the pier, as well as most of the houses, walls and cleitean. The cliffs appear to show a jointing pattern which developed as the rocks contracted on cooling. This should not be confused with the many minor dykes and sheets, consisting of a variety of rocks including rhyolite, dolerite and basalt, which cut the cliffs. These, the youngest St Kilda intrusions, are often softer than the surrounding rocks, and differential erosion has formed tunnels and arches, including the famous Tunnel at Gob na h-Airde[70,28,19] (Fig. 4.6).

The fact that so many of the St Kilda rocks are crushed and shattered is thought to indicate an explosive release of water late in the cooling process[20].

No Tertiary lavas survive on land in the Outer Hebrides but the very large numbers of Tertiary dykes could have been feeders for lava flows which have since been completely eroded[16]. Tertiary dykes, consisting of basalt and dolerite, occur throughout the Outer Hebrides, but are particularly numerous in South Harris and between Loch Shell and Loch Roag. The dykes are composed of crinanite (or olivine dolerite) and are typically 2–3m thick and may contain secondary zeolites (in amygdales) or chlorite, apatite, and iron ores. Olivine-free dolerites (which are tholeiitic) seem to occur only in South Harris, where they are best seen at Eilean Quidnish NG095861 and in Loch Grosebay NG157925 and may have amygdales 3–4cm in diameter containing brown glass or zeolites[65,5,21]. The dykes are predominantly oriented NW–SE, as would be expected from origins in the Cuillins centre of Skye (Harris dyke swarm) and Mull (Barra and Uists dyke swarm)[70,19].

The Shiant Islands are utterly different in appearance from the main islands of the Outer Hebrides, and it is not surprising to learn that they are more closely related geologically to Skye, being similar to the sills of the Trotternish area[20]. Most of the rock of the islands is crinanite, an olivine dolerite; the rock names change with increasing depth from crinanite through picrodolerite to picrite, reflecting (in broad terms) an increase in olivine content, as the heavy olivine would have sunk to the base of the magma prior to final cooling. Picrite and picrodolerite constitute much of the east coast of Garbh Eilean. The contact with the sediments which the igneous rocks intrude is marked by a thin layer of teschenite, a fine-grained olivine-free dolerite. There may be as many as four separate sills in the Shiants[20]. For most people the main interest of the geology of the Shiants will be the columnar jointing of the dolerite (Fig. 2.15). Some columns on the coast of Eilean an Tighe are broken at the base and are reminiscent of the Giant's Causeway in Antrim, but the towering columnar cliffs of Garbh Eilean and Eilean an Tighe make even Staffa, which is also known for its columnar dolerite, seem tame; these cliffs were memorably, if imaginatively, captured by William Daniell in his aquatints.

The three Madadh rocks at the entrance to Loch Maddy, together with nearby Ard nam Madadh NF951668 on mainland North Uist, are similar in composition to the Shiants, and display columnar jointing, albeit on a smaller scale[47]. The sill forming the Madadh rocks is perhaps more than 45m thick and lies on top of a 2.5m band of baked calcareous mudstones believed to be of Tertiary age which in turn lie on what may be another sill[21].

Stewart Angus. November 1991

Figure 2.15. Basalt columns, Eilean an Tighe, Shiant Islands

SEABED GEOLOGY

The British Geological Survey (BGS) embarked on a systematic study of the offshore geology of the UK Continental Shelf in 1966, and recently published a series of reports and 1:250,000 maps covering the whole of the area. The Minch is covered by *The geology of the Malin-Hebrides sea area*[24] and the seabed to the west of the Outer Hebrides by *The geology of the Hebrides and West Shetland shelves, and adjacent deep-water areas*[71]. Unless otherwise indicated these works are the source of all the material in this section.

There are obvious difficulties for the geologist attempting to work on rocks which may be covered by hundreds of metres of water. Fortunately there are now numerous boreholes drilled by BGS, and data from commercial drilling such as the BP *Ocean Alliance* exploration in mid-Minch in 1989 have been made available to BGS. There are also techniques such as high resolution seismic profiles (which may use explosives) and studies of gravity anomalies, which indicate the presence of buried rocks of different density to those around them. Unfortunately the area to the west of the Outer Hebrides is poorly known, with only five boreholes.

The islands of the Outer Hebrides and a great expanse of rock to the west form the Outer Hebrides platform, consisting of Lewisian gneiss. Though gneiss underlies the remainder of the area, it is rare on the surface of the Minch seabed to the east of the Minch fault, where it is overlain by younger rocks.

Downfaulting of sections of the gneiss has produced a series of sediment-ary basins (using the last word geologically rather than in the sense of deep water areas). Thus the Minch (east of the Minch Fault), the Barra Trough and the Flannan Trough all feature younger rocks.

Gravitational measurements have detected the seaward extension of the South Harris Igneous Complex west to the edge of the Flannan Trough. The Igneous Complex also occurs on the Flannan Ridge.

Torridonian sediments, which cover much of the gneiss of the north-western Scottish mainland, occur near the Western Isles only on the Rubha Reidh Ridge, as a mass roughly half-way between Kebock Head and the Summer Isles. These sandstones and conglomerates superficially resemble the sediments of the Stornoway Formation. This Torridonian deposit is linked to the mainland of Scotland by a ridge of Permo-Triassic rocks, probably the same formation as the Stornoway Formation, which also forms the bed of Broad Bay, much of the North Lewis Basin, and most of the Minch seabed east of Barra, and underlie most of the remainder of the Minch; to the west, they comprise much of the Flannan Trough and the Barra Trough.

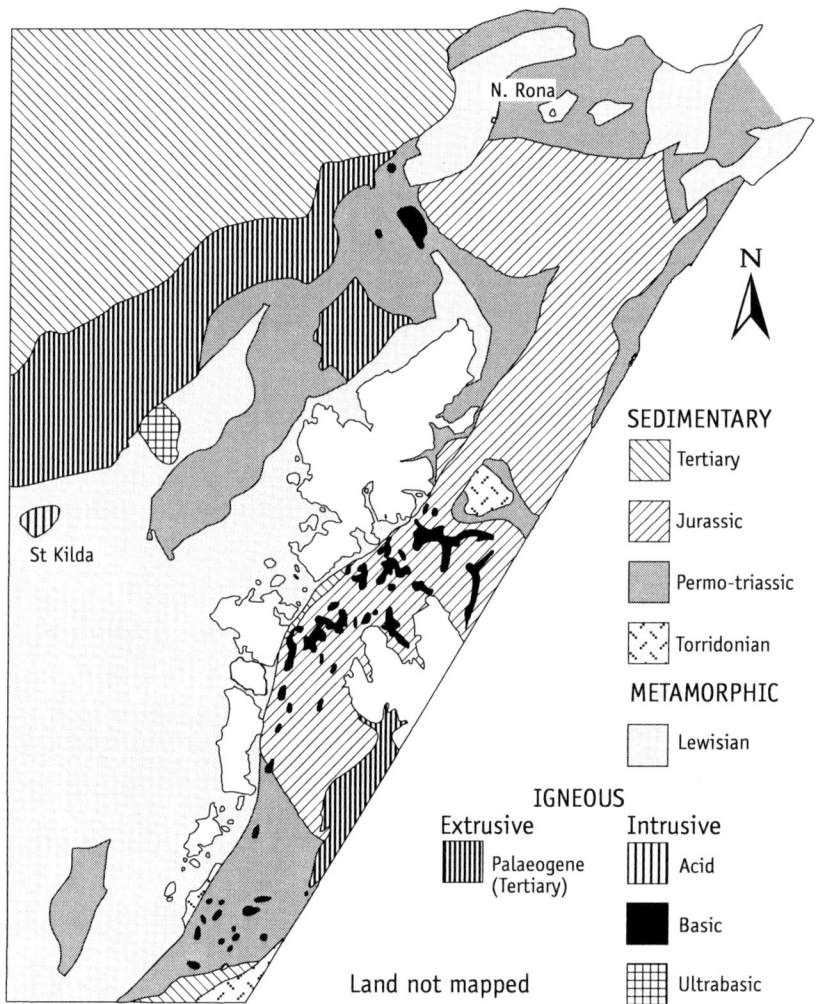

Figure 2.16. Seabed geology (after British Geological Survey 1991. Geology of the United Kingdom, Ireland and the adjacent Continental Shelf [1:1,000,000, North Sheet]. British Geological Survey, Keyworth). *Reproduced by permission of the British Geological Survey.*

There were several Tertiary igneous centres west of the Outer Hebrides associated with the opening of the Atlantic (see above) including St Kilda, Rockall, and the Anton Dohrn Seamount which rises steeply from the seabed of the Rockall Trough about 450km west of Benbecula. The Hebrides Seamount some 300km ESE of Barra and the Rosemary Bank, in the northern Rockall Trough, are also seamounts, but the latter is now partially buried by younger sediments. The Geikie igneous centre, around 80km north-west of St Kilda, is the closest of the Tertiary igneous centres and, like St Kilda, was probably a volcano. Other Tertiary igneous rocks probably associated with these igneous centres crop out over much of the western shelf, and also off north-western Lewis. These outcrops are often lavas and tuffs, so that they flowed over a surface exposed to the air. These igneous rocks represent a range of basalts, though most are olivine-tholeiites: as lavas and tuffs predominate, there must have been subsequent subsidence associated with the opening of the North Atlantic which placed them 3km below sea level.

The terrestrial expression of this Tertiary activity in the Minch is seen in the Shiants and in the Madadh rocks in North Uist, but much of the Minch to the south of the Shiants has scattered Tertiary sills, dykes, plugs, and vents, all detectable by gravity anomaly studies. In contrast, the northern Minch is virtually devoid of signs of Tertiary igneous activity, with one very notable exception: when aerial gravity measurements detected a straight line 110km long extending from Loch Maree in Wester Ross roughly NNW-wards beyond the Butt of Lewis. This 'Minch Anomaly' as it is now known, spawned the first ever BGS marine investigation in 1966. Surveys have revealed the structure to be a dyke of 1.1km width which does not even crop out on the seabed!

Where the Permo-Triassic rocks are overlain in the Minch, the younger (overlying) rock is for the most part Jurassic sediment. Westwards, these sediments lie beneath the west Flannan Basin and the West Lewis Basin. Jurassic rocks crop out on land in the Outer Hebrides only on the Shiants, but offshore the beds may be 1000-2500m thick, and the upper Jurassic rocks include Kimmeridgian shales, the main source rock for oil in the North Sea. A borehole 5km east of Raerinish Point yielded black Jurassic shales with ammonite fragments.

A tiny sedimentary basin just beyond the Minch fault off Rodel has been found by BGS drilling to consist of mudstones, containing carbonaceous fragments of plants and wood of late Oligocene age (c. 26–30 Ma).

Offshore sediments are described in the following chapter.

GEOLOGY AND MAN

In their paper on geology in the Symposium volume on the Outer Hebrides, Dr D.I. Smith and Dr D. J. Fettes[65] observed that "In reviewing the influence which the geology has exerted on the colonization of the Outer Hebrides by Man, it has to be admitted that on the whole the rocks have proved more of a hindrance than a help".

Certainly for the early settlers there was no shortage of stone for building materials for dwellings or structures such as Barpa Langass. The lack of sandstone or other bedded sedimentary rocks made building more difficult than, for example, in Orkney. The source quarry for the Calanais Stones is not known, but is said locally to be at Na Drommanan NB229336, some 1.5km NE of the main circle; the rock here is typical grey gneiss.

One of the most interesting stories of man and rocks in the Outer Hebrides is related in the BGS Memoir[21]. In 1965, the geologist J.S. Myers was walking in the Uig hills, and came across an old mill, complete with querns, at Loch Sandavat NB008303. What was unusual was that the millstones were made from appinite, a rock which had not been found in the Outer Hebrides. He passed on the news of his find to the British Geological Survey, who subsequently found an additional seven millstone locations within 3km of Carishader. Not only did they find the source quarry, and a new rock for the Outer Hebrides, but the quarry even contained a partially completed millstone! The quarry was located at the entrance to Loch Roag NB124318 and the dyke was subsequently traced for considerable distances. Another small appinite dyke was later identified on the foreshore near Shawbost NB268488.

Another partially completed millstone is said to exist at Griminish in North Uist[3], but my own attempts to locate this in 1981 were thwarted when I was informed that the site is now buried by blown sand. In her marvellous *West Over Sea* Pochin Mould[51] describes a mill in the river at Tamanavay, Uig, noting that the millstones were of a dark hornblende gneiss which were quarried from a special band which could be traced across to Loch Resort and the Clisham. The Ordnance Survey 1:10,000 map of the Monach Isles shows a 'quern quarry' at Port Mor NF642616. When encountered nowadays, many quernstones are broken.

The very hard Lewisian rocks of the Western Isles make good aggregate, and several quarries exist for this purpose, notably at Marybank, just south of Stornoway (Fig. 2.17); at Crogarry Beag in North Uist, and at Ceann an Ora, just north of Tarbert. Rusting quarrying equipment was left for many years in Glen Valtos in Uig, among other places. Interestingly, the 1:10,000 Ordnance Survey maps indicate the gneiss of the Marybank Quarry as 'conglomerate', a rare error on their part. Marybank Quarry was opened in 1939, and by the late

Stewart Angus, March 1988

Figure 2.17. Marybank Quarry, Stornoway

1980s had attained an annual output of about 130,000 tonnes, while Ceann an Ora in North Harris has an annual production of about 40,000 tonnes[64].

Hard though the Lewisian gneisses may be, they are thought to weather too easily for use as a building stone, and in the late 19th century, the stone used "in the best class of houses in Stornoway" was Torridonian sandstone from Isle Martin, at the mouth of Loch Broom in Wester Ross[50]. Lews Castle, however, is built from 'granite' from Bragar and Dalbeg[29].

The first record of the alleged mineral wealth of the Western Isles dates back to 1616, when an entry in the Privy Council said: "Ane patent ardanit to be past to Archibald Prymrois of the copper and lead mines of Ila, Sky and Lewis"[75].

The first mineral survey of the Western Isles was undertaken by the Reverend James Headrick in 1800, in a report to the then owner of Lewis, Lord Seaforth[31]. Perhaps his most important observation was that experiments could be devised to see if iron could be obtained from the 'bog-ore of iron', using peat as fuel in smelting. He noted that the Stornoway Formation was calcareous, and remarked on the stalactites of Gress and other sea caves. He also found "the most beautiful, regularly formed, and semi-transparent rhomboidal crystals of lime" at Kebock Head. Calcite may be associated with the mylonite here, as it is at Loch Carnan in South Uist[53]. Notes added in ink (probably much later) to the original version of Bald's 'Plan of the Island of Harris' of 1804[1], indicate: the feldspar, mica and quartz [pegmatite] at Northton; serpentine and indurated talc, lava and asbestos west of Finsbay Loch, and the 'Asbistos Quarry' east of the hamlet cluster at Borve.

In 1971, Robertson Research Associates presented the first of two reports[46] on the mineral potential of Lewis and Harris to the then Highlands and Islands Development Board.

They noted that there was no record of production of any metallic minerals in Lewis and Harris, though trials for iron had been made "to the north of St Clements in Rodel". The highest values of nickel (up to 600 parts per million) and chromium were associated with ultrabasic rocks while the highest levels of lead (up to 30ppm), zinc (350ppm) and copper (70ppm) were associated with shear zones (where one body of rock has moved relative to another). None of these were of economic importance.

They also noted radioactivity in the pegmatites of Harris, with readings of up to 25mR/hr, locally up to 50mR/hr, compared with 10-15mR/hr in acid rocks and only 2–3mR/hr in basic and ultrabasic rocks. The Northton pegmatite quarry on Chaipaval locally yielded a reading of 100mR/hr, while a red, rusty granite near Cliasmol (NB076066) gave 70–80mR/hr over a few square feet of rock. The radioactivity was thought to be due to thorium rather than uranium. The highest readings of all, 160-180mR/hr, were obtained from a granite pegmatite just south of Tarbert NG154998. [1 Rad corresponds to 0.01 sievert].

The 1971 report gave sufficient information on certain ultrabasic rocks, and on diatomite and kyanite, as well as the South Harris anorthosite, to warrant a more detailed report on these[45].

An ultrabasic rock at Loch Shell, cropping out from Rubha Cleit NB298084, through Loch Clachan Dearga, to Loch Cleit a'Ghuib Choille NB312097, contained up to 60% olivine. The original content had been in the region of 90%, but most of the crystals were partly converted to serpentine and accompanied by up to 5% green hornblende. The exposure was too small, dispersed, and inaccessible to be of economic value.

At Maaruig, ultrabasic rocks form cliffs up to 100ft high north of the village at NB202061. Though they cover a total area of 13.4ha, they are poorly exposed; outcrops have a distinctive rough, brown, weathered surface, while outcrops on the foreshore have a potholed type of weathering. Maaruig was not considered a suitable source of industrial olivine (used in foundries, for firebricks, and for heat reservoirs in storage heaters) due to difficulties in separating it from the rock and to the high iron content.

Diatomite is formed from the siliceous cells of fresh water plant plankton from late glacial or interglacial loch deposits. Because of its low density, porosity, large surface area, chemical inertness and lightness of colour, it has a wide range of uses including: a filtration medium, filler in paints, rubber, plastics, paper, insulation, absorbents, abrasives and as a catalyst carrier in chemical processes. It has been quarried on a small scale at North Tolsta and on Great Bernera, the latter in 1960.

The North Tolsta diatomite is centred on the basin of Loch Osabhat, which was drained in 1875. The deposit was up to 3m thick and 46m in diameter, giving a total reserve of 2,000 tons dry weight. A smaller deposit at North Shawbost NB262472 had a quarter of this. At a time when a deposit of 200,000 tons in Skye was of questionable viability, these deposits were thought to be of limited importance. Interestingly, Robertson Research abandoned further work after being told that both diatomite deposits were within nature reserves: neither is or ever has been within any site designated for nature conservation importance.

Between 1941 and 1945, quarries in the pegmatite vein on Chaipaval and in the Sletteval pegmatite operated by J.G.Gregory & Sons of Newcastle, Staffordshire employed sixty men and are said to have supplied most of Britain's feldspar requirements during the war[46,54]. Potassium feldspar from both sites was exported to the Potteries through Leverburgh for the manufacture of porcelain insulators[9]. Keith Nicholson[54] believed that the pegmatites of Chaipaval, Sletteval and Scarp deserved serious consideration for commercial exploitation of their feldspar, with the possibility of extracting uranium, niobium and tantalum as by-products.

Most of the second Robertson Research Report is devoted to the anorthosite deposit of Lingerabay. Anorthosite is very hard, and can be used as a reflective aggregate for road surfaces. It can also be used for abrasives, as a source of alumina (used in the glass industry), and as a whitener in cosmetics[2,65]. The white anorthosite can be used as a filler in plastics and paints, and as an abrasive in scouring powder and toothpaste. Following the successful use of anorthosite as edging for the A45 near Coventry in 1965, some 20,000 tons were removed from here between November 1966 and August 1967 by Kneeshaw Lupton Ltd, but only 10% of this was processed, the remainder being used locally for quarry roads. The operation finally closed in 1972 following problems with the quarry plant, extraction, and overburden (soil covering the deposit) which seems to cover most eventualities. During the flurry of interest in the 1960s, the anorthosite resource was thought to be in the region of 6 million tonnes[64], but at the time of writing Redland Aggregates Ltd are applying for planning consent to construct a superquarry here to extract 600 million tonnes, which would no doubt open new exposures for study, but which has profound environmental implications, the subject of a Public Local Inquiry (Fig. 2.18).

The anorthosite of Ness is also extensive in area, but lacks the comparative uniformity of the Harris rock, occurring intermittently among banded

Photo: P. & A. Macdonald. Montage: Turnbull Jeffrey Partnership

Figure 2.18. Computer visualisation of the proposed Lingerabay superquarry

gneisses and metasediments, and is yellowish-white[64].

Though some of the garnet-rich orthogneisses of the South Harris Complex contain up to 25% garnet (e.g. the NW slopes of Beinn Tharsuinn NG028881), there are difficulties in commercial-scale exploitation, as there are with the kyanite found in the metasediments of South Harris[65]. MacGillivray[43] described a garnet 10cm (4") in diameter in what appears from the description to be pegmatite on Roineabhal.

Beryllium, used in the nuclear and aerospace industries, has been found on Chaipaval, but not in commercial quantities[2].

The only known Scottish deposit of vermiculite, used as an expansion agent and lightweight filler, is at a small knoll near Rodel NG045834. This mica-like mineral is golden-yellow in appearance, but quantities are too small to be of any value[55].

The 'asbestos' of the Outer Hebrides does not occur in commercially viable quantities[2].

The best hope for successful mineral exploitation in the Western Isles probably lies offshore. The Jurassic rocks of the Minch Basins and basins west of the Hebrides are thought to include Kimmeridgian shales, which are often the source rock for oil.

GEOLOGICAL CONSERVATION

The rocks of the Outer Hebrides are of international significance: the study of the complex geology of these ancient rocks – which span more than half of geological time – is of great importance in deciphering the early history of the Earth[4]. It is vital, therefore, that these rocks be conserved. The most important geological sites in the Western Isles are scheduled as Sites of Special Scientific Interest (SSSI) by Scottish Natural Heritage. The Joint Nature Conservation Committee has already published some of the volumes of the *Geological Conservation Review* which will form a series of companion volumes to the biological *Nature Conservation Review*[61].

Fortunately, important geological sites are not so easily damaged as their biological counterparts. Even quarrying, which is popularly believed to be the greatest threat to geological conservation interests, may be beneficial, in that new exposures become available for study: only in extreme cases, where the rock has a very restricted distribution, or where the removal of an entire rock formation was contemplated, would quarrying be regarded as a serious threat. Road cuttings also open up new exposures and, like most quarries and quarry extensions, are regarded as beneficial forms of development by geologists[18].

Hard rock geological sites must usually, by their very nature, consist for the most part of bare rock, so that the designation is unlikely to attract the sort of criticism which has been noted for biological sites; some of the most important sites are on the foreshore, and are thus in 'no-man's-land'.

The scale of forestry in the Western Isles is such that it is unlikely to become a threat to geological interests by obscuring large-scale features. Likewise the fact that most of the salient features of the Lewisian rocks are large-scale means that few are likely to be threatened by hammering and over-collecting by geologists. The islands are sufficiently distant from most Universities to discourage large parties of geological students who tend to be the worst offenders in this field, though most are now observing the field code issued by the Geologists' Association. Most quarries and road cuttings in the islands have detached blocks at the base, and it is preferable to hammer these rather than the actual exposure. Some of the smaller, more interesting exposures such as Ardivachar Point, Orinsay (South Lochs) and Leanish Point (Barra), Cnoc a'Chapuill (Uig) and Loch Meurach (South Harris) should not be hammered. To avoid damage to vulnerable sites anyone wishing to hammer rocks in the Western Isles should first refer to the NCC's *Outer Hebrides: Localities of Geological and Geomorphological Interest*[4] or consult Scottish Natural Heritage. All rocks, even blocks at the base of quarries or cuttings, are private property, and the permission of the owner of the mineral rights is required before these can be removed.

SSSIs are not always the visitor magnets that some would have you imagine. Some, certainly, have spectacular features which, with the aid of interpretation, can both entertain and inform. Some of the more complex geological sites, however, can only be appreciated with considerable geological knowledge; they have been selected because they represent features of interest to research geologists rather than the public.

During the Lingerabay Public Local Inquiry, it was claimed that the quarry would make an excellent site for geological interpretation. One of the main attractions of the anorthosite for quarrying is its uniformity, and it is therefore ill-suited as a location for the presentation of the geological history of the Outer Hebrides. Having said this, South Harris as a whole has the most varied geology of any part of the island chain, and it would be easy to argue for some form of earth-science interpretation in the vicinity of Lingerabay. Indeed, the only interpretation of Outer Hebridean rocks in recent years was an exhibition mounted in An Clachan in Leverburgh by John Macaulay of Flodabay for the South Harris Historical Society.

That the rocks of the Outer Hebrides have a fascinating story to tell is hardly in doubt: the bare crags and man's exposure of new rock surfaces can reveal hitherto hidden details of the history of the development of the very crust of the Earth during the very early stages of its history.

Among the 42 volumes of the *Geological Conservation Review* will be one covering the Lewisian, which will no doubt synthesise the huge body of research that has been done in this field in the last few decades. As well as explaining why the sites presented for selection are important, it will put the Lewisian rocks in their wider context, giving the latest thinking on the way the gneisses and related rocks have thrown yet more new light on one of the biggest stories science has to tell. The Outer Hebrides have a special place in this story, and the publication of this volume should act as a stimulus not only for future research but for greater local interest in, and interpretation of, the very foundation of our islands.

3

GLACIATION

―――⇒•●•⇐―――

*I was taking down a story from a man, describing how twin giants detached a huge
stone from the parent rock, and how the two carried the enormous block of many tons
upon their broad shoulders to lay it over a deep gully in order that their white-maned
steeds might cross. Their enemy, however, came upon them in the night-time when
thus engaged, and threw a magic mist around them, lessening their strength and
causing them to fail beneath their burden. In the midst of the graphic description, the
grandson of the narrator, himself an aspirant teacher, called out in tones of superior
authority, 'Grandfather, the teacher says that you ought to be placed upon the stool
for your lying Gaelic stories.' The old man stopped and gasped in pained surprise. It
required time and sympathy to soothe his feelings and to obtain the rest of the tale,
which was wise, beautiful, and poetic, for the big, strong giants were Frost and Ice,
and their subtle enemy was Thaw. The enormous stone torn from the parent rock is
called 'Clach Mhor Leum nan Caorach', the big stone of the leap of the sheep. Truly
'a little learning is a dangerous thing'! This myth was afterwards appreciated by the
Royal Society of Edinburgh.* (Carmichael 1928)[11].*

Though most of the large-scale features of the landscape of the Outer Hebrides
existed prior to the action of ice-sheets and glaciers, the fine detail, and the
overall 'flavour' of the landforms of the islands are substantially a legacy of
glaciation. The interaction of glaciers and ice-sheets with the pre-existing
topography, together with the deposition associated with melting, is responsible
for the development of the glens and corries of the upland areas, for the scoured
and pitted platforms of northern Lewis and North Uist, for the excavation of the
deep basins of the larger inland lochs and coastal fjords and even, indirectly, for
the formation of machair.

The story of how the history of glacial events affecting the Outer
Hebrides was decoded is a fascinating one: in the 1970s, a young student from
Aberdeen University, Jost von Weymarn, debunked a theory which had been
almost universally accepted for a hundred years. I was lucky enough to be able
to accompany von Weymarn on some of the fieldwork which led him to his new
theory of island glaciation – which is now as universally accepted as its

predecessor was. Though many important questions remain unanswered, and experts disagree on the interpretation of some of the available evidence, even an amateur armed with the wisdom of hindsight may now see how marvellously von Weymarn's theory fits the field evidence, and will no doubt wonder how it took so long to get it right.

The story begins many years before the onset of glacial conditions.

BEFORE THE ICE

As described in the previous chapter, some of the rocks now exposed on the surface of the Western Isles were formed at depths of 30 or even 40km in the Earth's crust below other, unknown, rocks. If the existence of rock 40,000m (25 miles) above our present land surface is difficult to envisage, then the removal of such a thick layer must be even harder to imagine. Clearly, earth movements must have raised these rocks (uplift) while other processes must have combined to remove the overlying layers.

During the Tertiary (see Chapter 2) the warm, moist subtropical climate which prevailed in the Outer Hebrides encouraged the chemical breakdown or *weathering* of certain rocks such as dolerite and granite, to the extent that the resulting 'rotten' rock can be prodded with a finger, sometimes to a depth of 5–6m. Such deeply decomposed bedrock occurs in a belt running from Aline Lodge (NB197118) to Carishader in Uig; between Tarbert and Scalpay; and between Caversta and Gravir in South Lochs[52]. Examples may be seen near Loch Airidh na Lic, Stornoway; at Breaclete, Great Bernera[22] and by the side of the road just south of Carishader, Uig (NB103318). Decomposed bedrock also occurs in North Uist, and a fine example, weathered to a depth of 6m, may be seen at the head of Loch Eport (NF833634)[57]. Postglacial weathering is measured in centimetres rather than metres, so it must be assumed that almost all of this decomposition predates glaciation. Such deep weathering has also been noted in Skye and Rum but is particularly well developed in NE Scotland around Buchan[26,25].

The question is: how did these soft rocks survive glaciation? Though it has been suggested that they could have become saturated with water frozen solid[22] it is more likely that they were sheltered from the worst scouring by neighbouring crags of hard rock or were passed over only by 'cold based' ice sliding without much pressure[41].

Deeply weathered rocks are uncommon in the Western Isles because of almost universal ice scouring, and it is reasonable to conclude that prior to the onset of glaciation there were many such areas, which would have been easily removed as the glaciers and ice sheets advanced[79].

Even if only the maximum depth of rock (40km) is considered, and it is assumed (for ease of calculation) that denudation took place over a period of 2,000 million years (Ma), an erosion rate of 20m every million years is required, equivalent to 1mm every 50 years, which seems much easier to believe. Erosion can progress much more rapidly than this, and it is believed that between 58 and 56 Ma some 2km thickness of lava was removed from Mull[27].

Softer rocks would obviously have been eroded more quickly, leaving the harder ones standing out as hills or even mountains. Examples of hard rock masses which have eroded more slowly than the surrounding countryside include the Corodale gneisses of Beinn Mhor and Hecla in South Uist; the South Harris Igneous Complex rocks forming Chaipaval, Bleabhal and Roineabhal; the granites of the Uig hills, North Harris and northern South Harris; and the pseudotachylite of the North and South Lees and Eaval. In contrast, the softer metasediments of South Harris, for example, were eroded more quickly than the surrounding rocks, permitting the initial formation of the basin now occupied by Loch Langavat[71]. On the other hand, the isolated hills which rise above the plateau of northern Lewis are not so easily explained, as most of them seem to be geologically identical to the rocks of the surrounding platform[22]. The areas most vulnerable to erosion would probably have been those where earth movements or igneous activity had weakened the rocks, either structurally or chemically[85].

Rock removal or *denudation* involves both *weathering*, the processes which involve the disintegration of rock, and the exposure of new layers to these processes by *erosion* by wind, water and gravity. Weathering may be physical, involving temperature changes, plant growth and frost action, or chemical, where minerals react with atmospheric gases or water. When the temperature of water falls below 4°C (but especially below 0°C) it expands in volume by up to 9%. The freezing of water which permeated cracks in the rocks would have exerted a strong disruptive effect, eventually breaking off sections of rock. Frost-shattering caused by alternate freezing and thawing is responsible for adding to the enormous scree beds on mountains such as Sron Scourst (NB1009) and Oreval (NB0809) in North Harris[7].

That huge areas of northern Lewis are at a very similar level is self-evident to anyone who has viewed the area from any height, but a general uniformity of altitude is also visible elsewhere in the islands, even in the summits of the uplands of southern Lewis and North Harris. This might be dismissed as coincidence were it not for the fact that comparable uniformity is visible over large areas of the Highlands. It had been suggested that each uniform level represents an ancient erosion surface of marine or perhaps terrestrial origin formed during the Tertiary Period, though the role of harder and softer rocks would obviously complicate the picture[49,22,21,30,40]. Recent interpretations suggest that these surfaces were exposed to the air, and erosion occurred as a process of 'etching' involving chemical weathering of the rock and the removal of the erosion product by a combination of gravity and streams[27].

Whether or not these Tertiary 'planation surfaces' are genuine, it is certain that the main landform pattern was established prior to the beginning of glaciation: ancient weathering and erosion had already identified and excavated the main valleys; the ice acted on this template.

THE COMING OF THE ICE

Though glaciations are known to have occurred during the Permian and Carboniferous Periods (230–345 million years ago), and even during the Precambrian (>600 Ma)[38], the glaciations which had the most profound (known) effects on the north of Scotland were those of the Quaternary. Although the Quaternary is formally defined as the last 1.8 million years[9], the first known signs of glaciation in the area have been dated at around 2.4 Ma by the Deep Sea Drilling Project near Rockall[65]. The cores extracted by this research programme show a fluctuating climate over the last 2.4 Ma, with many colder and warmer periods.

The extreme north of Europe and the Alps still support modern glaciers, and a drop in the average summer temperatures of only a few degrees would bring such conditions back to Britain[46]. The Ben Nevis range and the Cairngorms sometimes carry 'permanent' snow beds for long periods, only to lose them in exceptionally mild years.

When winter snow fall exceeds the amount of snow lost by summer melting, and this continues for a number of years, permanent snow beds build up in dark gullies or on the sides of valleys. Pre-Quaternary erosion would have created many such valleys and gullies where snow and ice would have accumulated.

GLACIAL EROSION

Glaciers erode and excavate material by a combination of processes: embedded debris in the base of the ice abrades (scratches) the underlying rock and scrapes away soil, while chunks of rock are 'plucked' from the rock at the base and sides of the moving ice. Indeed, under a great weight of ice and the rocks embedded therein, the glacier sole would have exerted a chisel-like action on the bedrock. Plucking functions by pressure melting on the 'up' side of rock obstacles and refreezing on the 'down' side, and the fractured rock is removed; this is particularly well illustrated by roches mouttonées (see below). The extent of erosion is dependent on such factors as the weight of overlying ice, the temperature at the base, the amount of embedded material, speed of ice movement and, of course, the nature of the substrate[73,31].

Assuming that the prevailing winds were then, as now, from the south-west, drifting snow would have built up in any sheltered NE-facing depressions and valley-heads, out of the sun. Under freeze-thaw conditions, the wall of the depression would have been steepened while its floor would have been excavated by the movement of the ice, formed from the compaction of successive layers of snow. Thus depressions would have been deepened and steepened, forming a characteristic armchair-shaped depression internationally termed a *cirque* but known in Scotland as a *corrie*, from the Gaelic *coire* for such a feature – Dwelly, in his famous dictionary, points out that the word should be pronounced differently from *coire* meaning 'kettle'. Most of our corries are on the north and north-east sides of mountains, away from the sun, and sheltered from the prevailing south-westerly winds[22]. This situation and aspect also mean that any snow blown from the summit areas by winds from the south and west would accumulate in the corrie[9]. It is also interesting to note that the altitude of corries is lower in NW Scotland than in NE Scotland, possibly reflecting the increased snow cover which would be associated with moist, westerly winds[9]. Coire Roineabhail, South Harris (NG0486) (Fig. 3.1) and Coire Dibidale in Uig

Stewart Angus, October 1984

Figure 3.1. Aerial view of Coire Roineabhail from the north

(NB0424) are particularly fine examples, and there is an interesting pair of corries at two different levels on the north side of the Clisham range. There are several others in the mountains of North Harris and in the Beinn Mhor-Hecla range of South Uist.

Corries are too well-developed to have been produced by the last glacial period alone, and it is likely that they formed over a large number of cold periods, some of which were not long or cold enough to produce larger glaciers or ice sheets (see below)[9].

With the onset of even colder conditions, the growing mass of ice in the corrie-glacier would have begun to move downwards under the influence of gravity, growing into a valley glacier. Moving very slowly down the valleys, the glaciers continued to erode all they touched.

Even if the exploitation of natural lines of weakness is included, these processes alone are not really sufficient to explain the great depths of some valleys and glacial troughs – Loch Suainaval, for example, is 65.7m deep – and it has been argued that glacial erosion was assisted by another process which can be seen in some quarries today. As the glacier removed rock by abrasion and plucking, the relief from the weight of rock removed would have caused expansion of the rock beneath, causing further shattering. This process would have been most severe where other glacial processes were most active, so that the deepest valleys and basins would have become deeper still[42,44,66].

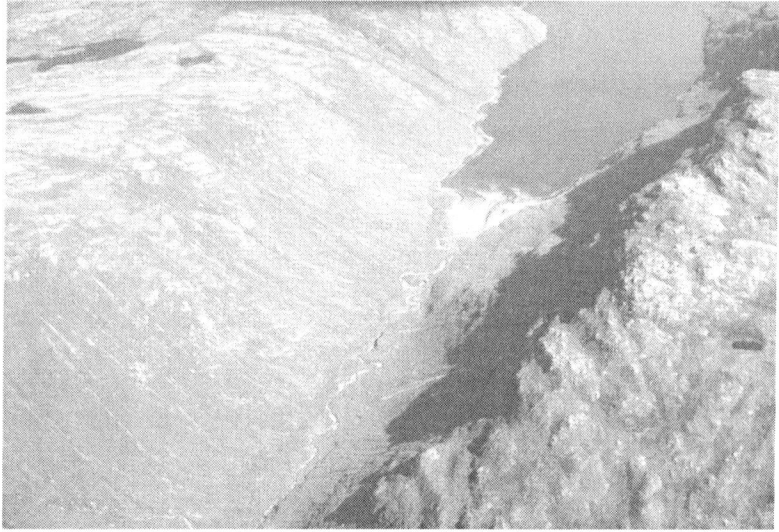

Stewart Angus

Figure 3.2. Gleann Claidh, Pairc, a typical glaciated glen

Over a long period, each valley glacier excavated the sides and base of its valley to produce the characteristic profile, usually referred to as a 'U-shaped valley', though the sides of the U are rarely, if ever, so steep in Scotland as they are in North America or New Zealand, and the profile of Scottish valleys is really more parabola-shaped[31]. Small valley glaciers (with comparatively shallow channels) fed large valley glaciers (with deep channels) so that many of the larger glaciated valleys have smaller tributary 'hanging' valleys. A classic, if simple, glacial trough may be seen at Gleann Claidh, in Pairc (NB2407) (Fig. 3.2); though very remote, this valley is a conspicuous landmark on the air route between Stornoway and Benbecula. All the larger valleys of the Outer Hebrides have this characteristic parabolic shape except for Glen Valtos in Uig, though the profile is not always as obvious as it is in Gleann Claidh. Glen Valtos will be explained below.

It is easiest to refer to glaciated valleys as 'glacial troughs', as this term embraces not only the typical glens we see today, but also the deep, excavated basins now occupied by dark, steep-sided fresh water lochs such as Loch Suainaval, as well as those now submerged as the 'fjords' which now form many of the larger sea lochs. Pedants of place-name etymology would no doubt have me point out at this stage that the real name of Loch Suainaval should be 'Suainavat' so that the last syllable means 'water' rather than 'mountain', but I fear that a few mistakes by the map-makers have already triumphed over accuracy.

As the valley glaciers met the sea, the higher temperatures there would have caused melting, and the glaciers would also have 'calved' into icebergs, so that the erosive power of the glaciers decreased seawards. Thus, the seaward ends of these troughs are shallower than their middle sections. It is important to stress that the ridge or *sill* near the mouth of a genuine fjord is composed of bedrock rather than glacier-borne debris. The largest glacial troughs are now submerged as some of our major sea lochs, one of the largest being the outer arm of Loch Seaforth. East Loch Tarbert and Lochs Seaforth (inner and outer), Shell, Claidh and Tamanavay are true fjords, all possessing both glacial troughs and sills[81]. Loch Eport lacks a deep basin and sill and is thus not a fjord: its origins are thought to be fault-related[57]. Any rise in sea level of more than a few metres would transform Loch a'Ghlinne in North Harris into a fjord.

Gleann Chliostair in North Harris is a typical glacial trough, showing glacial over-deepening at Loch Chliostair (now impounded by a dam) and features a hanging valley south-west of Ullaval and a corrie at Loch Maolaig (Fig.3.3). The adjacent glen has hanging valleys at Caadale Ear (NB091117), Glen Uisletter (NB1109), Glen Stuladale and the valley of Abhainn a'Chlair Bhig, with corries and lochans at the head of the last two.

Stewart Angus, September 1984

Figure 3.3. Gleann Chliostair, North Harris, with the corrie loch of Loch Maolaig on the left

With continuing deterioration in climatic conditions, even the larger valleys were unable to cope with the accumulation of ice and snow, and the ice spilled over the valleys, forming a local ice cap which gradually expanded into an ice-sheet.

Glacier ice presently covers 10% of the Earth's land surface but, at its maximum, the Quaternary ice covered 31% of the land[86]. As the ice-sheet moved across the Hebridean land surface it exploited all the available lines of weakness: geological fault lines, the softer rocks of minor igneous intrusions, and the lines of weakness above buried intrusions are likely to have been followed by the ice[85], in most cases because pre-glacial erosion had already excavated these lines to form depressions or valleys. The ice-sheets left their own landforms as they passed, many of which can now be used to deduce the direction of ice flow. Sometimes the directional flow was so strong that the ice crossed the lines of weakness rather than flowing along them, often leaving distinctive traces as it did so.

With the exception of the inland part of northern Lewis, which is an ice-scoured platform now covered with till and peat, scratches or *striae* on polished rock surfaces reveal the direction of flow of the ice-sheet or glacier. These striae are best seen on rock recently freed from overlying soils, and are most reliable where they cross the 'grain' of the rock[81].

Where the ice-sheet met an outcrop of very hard, resistant rock, it was forced over and around it, sometimes depositing a 'tail' of debris in the 'lee' of the outcrop. The finest examples of these *crag-and-tail* formations in the Western Isles are probably those south of Loch Valtos on the northern shores of Loch Erisort (NB310208), plainly visible from the main road (Fig. 3.4). There are others in the Breasclete area and around Uig Sands, but they become fainter northwards, and are absent north of a line between Stornoway and Carloway[81,82]. There is a single example in South Harris, at Rubha Romagi, near Horgabost (NG035963) but the 'tail' has been excavated for road works[81,52]. The countryside around crag-and-tail features often has large-scale streamlining of rock and till[52]. Though it had been suggested that the North and South Lees of North Uist were gigantic crag-and-tail features[15] this possibility was dismissed by Ritchie[57].

The typical landscape of the Lewisian gneiss country has been described as 'knock-and-lochan' (more properly 'cnoc and lochan') topography because of the irregularly distributed rocky knolls and lochans which characterise the area[43]. Other terms used to describe this terrain which is typical of many parts of northern Lewis, South Harris, and North Uist are 'glacial roughening'[66] and

Stewart Angus, June 1977

Figure 3.4. Crag and tail formations at Laxay, Lewis. The direction of ice movement was from right to left (north-eastwards).

'areal scouring'[23]. Many of the depressions are now filled with peat rather than water.

Some of the rocky knolls of this landscape will be seen to be more regularly shaped than others, being perfectly described by the term *whaleback*. Others are similar, but 'quarried' at one end, being termed *roches moutonnées* (Figs 3.5 and 3.6). Whalebacks and roches moutonnées are notoriously variable in form, some being completely smooth, and others face a direction other than that of ice flow due to local variations in rock structure[23]. Fortunately they tend to occur in swarms, and the direction of ice flow can usually be determined from the examination of a number of perfect whalebacks, ideally in conjunction with other features in the vicinity, such as striae. These landforms are abundant in many parts of the Western Isles but are particularly well-represented in eastern South Harris[81].

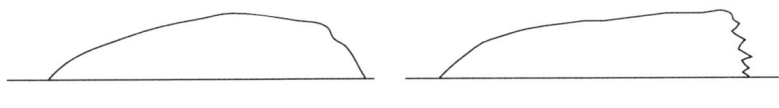

Figure 3.5. Whaleback (left) and roche moutonnée. The latter is 'quarried' by the passage of the ice.

Stewart Angus, December 1996

Figure 3.6. Whalebacks at Finsbay, South Harris

GLACIAL DEPOSITION

The base of a moving ice-sheet may carry huge quantities of erosion products which themselves may erode and be eroded as they travel. As it moves, the debris-laden base of an active ice-sheet may be subjected to varied loading and stress, so that instead of eroding the substrate, it may 'plaster' it with successive layers of debris, building up a deposit of *lodgement till,* consisting of an assortment of rock particles ranging in size from rock flour to huge boulders. Most of the till (also known as boulder clay) of the Western Isles is lodgement till, but locally there are deposits of *flow-tills*[52] which, as the name suggests, have moved in a liquid or semi-liquid form.

Till deposits up to 6m thick and overlain by peat cover most of northern Lewis, but elsewhere in the islands till tends to occur in isolated pockets, often in the 'lee' of rocky knolls. At Traigh na h-Uidhe, near Northton, there are till beds 10m thick[52].

Till associated with the ice-sheet was usually deposited as an amorphous layer or sheet, but sometimes it was laid down in the form of *moraine.* Moraines occur in mounds 2-5m high, and are particularly common in the Uists, but also occur in Harris and south-west Lewis[52]. Conspicuous examples may be seen around Samala, Baleshare. These moraines seem to have been deposited from the base of a moving ice-sheet, and occasionally resembles *drumlins*[52] (stream-lined mounds aligned in the direction of ice movement). Remains of rock-cored drumlin-like mounds may be seen between North Uist and Baleshare, where they have been eroded by rising sea levels[15], while a quarried mound at Lochportain NF950723 may indeed be a drumlin[52].

As will be explained below, the last ice-sheet glaciation which left the layers of till was followed by a local glaciation which was confined to corries and valleys. Instead of till sheets, the valley glaciation left behind a more distinct series of landforms collectively known as *morainic drift*[52]. The most distinctive of these are probably the *hummocky moraines* which resemble the moraines left by the ice-sheet. These may have formed by the deposition of debris from the surface of rapidly melting valley glaciers or at the glacier front[8,84,5]. Hummocky moraines are frequent in the valleys of Uig and North Harris, and fine examples may be seen just north of Loch Voshimid (NB105138) (Fig. 3.7)[52]. Hummocky moraines have also been recorded in the Pairc area of Lewis [82] and at Loch a'Choire (NF817357) on the northern slopes of Hecla, South Uist[52]. Recent work in Skye and the NW Highlands[4,5] has linked different types of hummocky moraines to different patterns of ice decay. Detailed mapping of the Outer Hebrides hummocky moraines is required to interpret their formation.

Some of the hummocks of the morainic drift form linear ridges or *flutes* aligned in the direction of ice flow, though this linearity may sometimes be

Stewart Angus, October 1984

Figure 3.7. Hummocky moraines north of Loch Voshimid, North Harris

difficult to discern from ground level, and they are best seen from adjacent high ground or from the air. The best examples of these *fluted moraines* are probably those just north of Loch a'Sgail (NB132084) and in Glen Uisletter (NB1109), both in North Harris[52]. Fluted moraines often occur on higher ground than hummocky moraines, sometimes on fairly steep slopes at the heads of glacial troughs, so that it is likely that flutes were deposited by rapidly-moving ice[68]. There also fluted moraines at Coire Roineabhail (NG047863)[52], indicating that a local valley glaciation also took place in South Harris.

Transverse moraines, aligned at right angles to the direction of ice-flow, may be seen on the slopes to the east of Loch Langavat, North Harris (NB1614, NB1615). Other examples occur on the west side of Glen Meavaig (five ridges at NB102105), in the valley of the Abhainn Cheothadail, near Eishken (NB305125) and on the slopes east of Loch Ulladale (NB083143). Some of these ridges have fluted surfaces. The transverse moraines of Lewis and Harris seem to have formed at glacier snouts[52].

The composition of a till is obviously determined by its origins, and if the contents of till can be traced back to a particular locality, the direction of ice flow can be deduced. The colour of a till may also yield clues to the direction of ice travel. Till derived from Lewisian gneiss is typically light green in colour, whilst that derived from the Stornoway Formation is purplish-red. Examination of the tills at the margins of the Stornoway Formation should, therefore, tell us the direction in which the ice travelled over these areas[81].

Pebbles or boulders consisting of a rock type other than that underlying the till (ie having been transported by ice) are termed *erratics*. The origin of the erratics, and therefore the direction of ice flow, can often be determined fairly accurately. Torridonian erratics are frequent in the tills of the extreme north of Lewis, at Tolsta Head and, to a lesser extent, in north-western Lewis and the Eye Peninsula. They are rare south of Carloway or Stornoway[17,82]. Torridonian erratics occur sporadically on the beaches of the west coast from South Harris to South Uist, and are frequent in Barra and the adjacent islands, where they are often mixed with chalk flints[34,35,36,17,69]. Jurassic limestones have been found at Tolsta Head and in north-east Lewis, Carboniferous limestones on the west and south coasts of South Uist, and Moine rocks in Barra and Vatersay[52]. As none of these rocks occur in the Western Isles, they must have been transported from the east by ice moving in a westerly direction, either from the north-west mainland, or over the bed of the Minch. Though Mesozoic rocks are common in the Minch (see Chapter 2), no Mesozoic erratics have yet been identified in the Outer Hebrides; erratics of Skye rock types are also absent.

Sometimes the smaller fragments of till are washed away, leaving the large boulders behind. Such boulders may litter some hillsides while others – often erratics – are perched conspicuously where the retreating ice has dumped them, looking almost as though someone had put them there (Fig. 3.8). There

Stewart Angus, July 1996

Figure 3.8. Perched blocks on skyline, north of Mealisval, Uig, Lewis

are spectacular perched blocks near Manish School in Harris (NG109895) and above Northbay, Barra (NF701045). The Northbay block is thought to have a volume of 670m³ (24,000ft³)[20].

A number of beds containing fragments of marine mollusc shells occur in coastal till and other deposits between Port Skigersta and Dell Sands (NB485624), at Garrabost (NB507333) and Tolsta (NB545470)[20,2,81,52]. The Port of Ness beds contain only cold-water species while those between Traigh Sanda (Eoropie) and Dell Sands include cold and warm-water species. Though this suggests that the beds were deposited at different times, it is possible that a complex series of events involving an ice-dammed loch brought about the deposition of both groups at around the same time[52]. The shell beds of northern Lewis are very complex, however, involving contradictory features, and much more work is required in this area before the detailed glacial history can be accurately determined. Radiocarbon dating suggests that the shells in the beds of north-west Lewis are between 34,000 and 40,000 years old, while those of Garrabost are approximately 23-26,000 years old[80]. The remains of kelp embedded in till clays have been reported from Garrabost[20].

The evidence of till colour, marine beds and erratics has to be treated with caution, as it is always possible that they may have been moved after their original deposition – e.g. by slope processes such as soil creep, by a subsequent glaciation, or transport by meltwater. This problem can to a certain extent be overcome by the specialised technique of 'till fabric analysis', coverage of which is beyond the scope of this work.

Our finest examples of fluvioglacial landforms – those produced by meltwater during glacial retreat – occur in the Carnish area of Uig. The largest of these feature is the Carnish *terrace*, a flat-topped ridge some 60m high at Druim Carnish (NB032316) consisting of sands and gravels of Lewisian origin. This terrace, often erroneously described as an *esker* (see below), was probably laid down in a meltwater loch impounded by ice to the west. A similar feature at Crowlista (NB038336) may have been part of the Carnish terrace before the deposits were breached by the sea, but it could have been entirely separate[81,52].

There is a large deposit of moundy and terraced gravel to the south of Carnish and, as at the Carnish terrace, the surface of this deposit is marked by the presence of *kettle-holes*[52], formed by the melting of ice blocks within the sediment. Most of the kettle-holes are filled with water but one or two are dry, and rocks which would have been embedded in the stagnant ice may now be seen at the bottom of the crater-like holes. Kettled boulder drift occurs to the east of Loch nan Eang, in the Clisham range[51]. Other terraces of sand and gravel occur near the mouth of Gleann Sandig, near the western shores of Loch Langavat (NB1517), immediately north of Kerasclett Beag, near Kinlochresort (NB110151)[52] and on the banks of the Heidagul River in Carloway (NB219428)[19].

There is a small *esker* – a ridge of gravel laid down by a meltwater river in, on or beneath the ice – with hummocky gravels near Loch Langavat (NB200210) and another on the slopes between Beinn a'Bhoth and Loch Langavat (NB141173). The ridges to the south and west of Loch Scaslavat, Uig (NB0231) may also be eskers[52].

Other fluvioglacial deposits have been identified between Habost and Cladach Cuiashader in Ness, and between Ardroil and Uig Lodge in Uig[52].

Meltwater has not only been responsible for deposition, but also for erosion. The gorge through which the Abhainn Caslavat flows at NB037294 is a meltwater channel, but there is a much better example at Glen Valtos, which stretches between Timsgarry and Miavaig in Uig, a distance of some 2.5km. Here, a meltwater river beneath a glacier has eroded a spectacular V-shaped valley in the hard Lewisian rocks to a depth of up to 70m. Interestingly, the meltwater flowed from west to east, the opposite direction from modern (east to west) water flow, and it is believed that there must have been thicker ice offshore at this time, possibly with an ice dam over what is now Camas Uig, but the history of glacial decay in this area, which would explain these anomalies more clearly, is not yet properly understood[76].

THE PATTERN OF ICE MOVEMENT

In 1873, James Geikie, brother of the famous geologist Archibald Geikie, presented his theory of the glaciation of the Western Isles, which was accepted for the next one hundred years. His interpretation of whalebacks and striae, and the presence of mainland erratics in certain areas led him to believe that the Western Isles had been glaciated by a large mainland ice-sheet which had moved westwards over the islands[19,20], even though the mineralogist W F Heddle had previously identified evidence of eastward ice movement near Tarbert[48]. J F Campbell[10] disputed Geikie's claim of 1873, suggesting that "the ice came from NNW through a gorge at Tarbert", but his protests were ignored. Even the great Jehu and Craig[34-37] did not dispute Geikie's conclusions, though they themselves had found evidence of the eastward movement of erratics in North Uist[35].

In the early 1970s, Jost von Weymarn, a postgraduate student at Aberdeen University, re-examined the glacial phenomena of Lewis and Harris and, from the orientation of striae, whalebacks, and crag-and-tail formations, together with the red discolouration of the Eye Peninsula tills, reached the (then) surprising conclusion that the last ice had moved out radially over Lewis and Harris from three major centres over Loch Langavat, West Loch Tarbert, and Pairc[81,82]. Working independently, Derek Flinn[17] came to a similar conclusion, providing evidence for local glaciation over much of the Western Isles archipelago (Fig. 3.9).

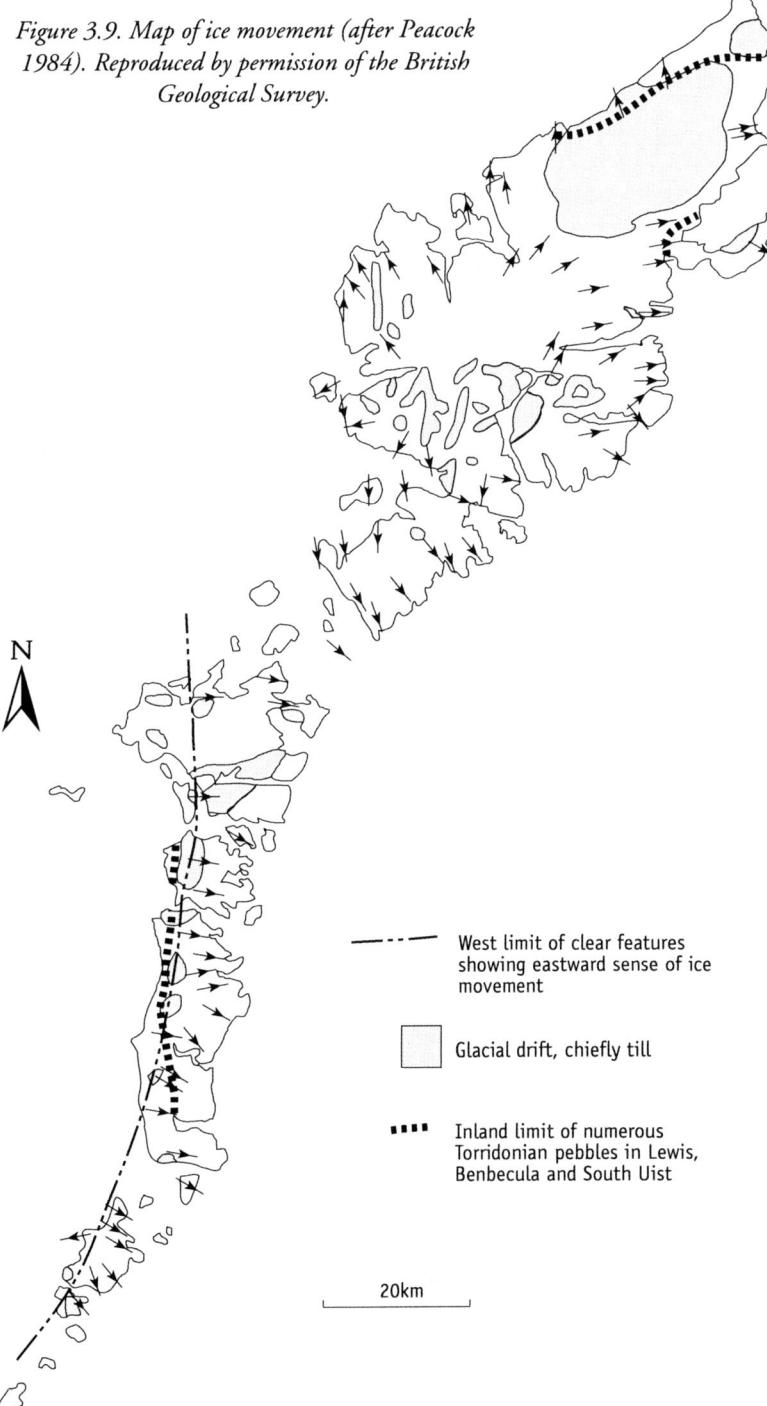

Figure 3.9. Map of ice movement (after Peacock 1984). Reproduced by permission of the British Geological Survey.

N

West limit of clear features showing eastward sense of ice movement

Glacial drift, chiefly till

Inland limit of numerous Torridonian pebbles in Lewis, Benbecula and South Uist

20km

Erratics of western gneisses in eastern South Uist indicate not only that ice moved eastwards there during the Hebridean ice-sheet glaciation but that the *ice-shed* (the glacial equivalent of a watershed) was situated to the west of the modern watershed[13].

Nevertheless, the presence of marine beds and mainland or Minch erratics implies that the islands were glaciated from the east. There is therefore incontrovertible evidence that the islands of the Outer Hebrides have been glaciated from local centres *and* from the east.

The most likely explanation is that there have been at least two separate glaciations. The earlier involved an ice-sheet which crossed the Minch, carrying with it Torridonian and other rocks which were deposited as erratics. This ice-sheet must have been at least 1500m thick over Loch Broom in order to have reached Lewis[82]. The later glacial phase involved a Hebridean ice-sheet, with ice radiating out from centres in North Harris and southern Lewis[82] and from an ice-shed lying slightly to the west of the present watershed in the southern islands (Fig. 3.9)[17]. This glaciation redistributed many of the tills and erratics of the previous one, e.g. in north-west Lewis[81]. Evidence for this pattern can be seen at Galson in north-west Lewis, where tills of eastern origin containing Torridonian erratics are overlain by tills of local origin[81]. There is a similar example at Cliad in Barra[50].

It may or may not be important that there are no erratics on the east coasts of the Uists and Barra: perhaps the ice carried them straight over to the west coast. The presence of erratics and shell beds confirms only that ice from the Minch reached the Outer Hebrides: there is no unambiguous evidence that an ice sheet from the Scottish *mainland* reached the Outer Hebrides during the Quaternary[28].

The extreme north of Lewis and the eastern part of Tolsta Head are some distance from the local centres of glaciation, and there is evidence that local ice did not reach these areas, and that the last ice there was of mainland (or Minch) origin (Fig. 3.10). While it is likely that the 'mainland' glaciation of Ness and Tolsta was contemporaneous with the local glaciation, this has not actually been established[82]. The northern limit of local ice was Breivig near Melbost Borve NB413582, where a drift ridge is discernible for a distance of over 6 km; the existence of an identified terrestrial ice limit of this period is unique in Scotland, and the sequence of deposits in northern Lewis makes this a very important study area, particularly as it has been suggested that the extreme north of Lewis could have been free of ice during the last glaciation of Lewis[80,25]. Recent work on the Galson beach, however, based on the presence of thin glacial deposits above the beach layer and on the presence of deposits of till and on striae on higher ground in the immediate vicinity, strongly suggests that there was no ice-free area in northern Lewis during the Late Devensian glacial maximum. This

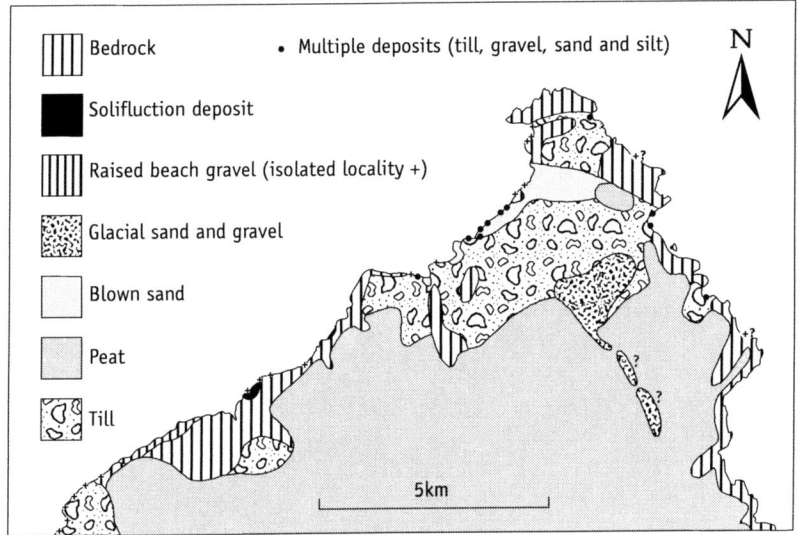

Bedrock

Solifluction deposit

Raised beach gravel (isolated locality +)

Glacial sand and gravel

Blown sand

Peat

Till

Multiple deposits (till, gravel, sand and silt)

N

5km

Figure 3.10. Quaternary deposits and peat in northern Lewis (after Peacock 1984). Reproduced by permission of the British Geological Survey.

is supported by other recent studies[72,53] which place the last ice limit some distance offshore. The features around Breivig represent a retreat stage of the ice sheet. It is possible that the Galson beach survived this glaciation by virtue of its position on the boundary between the local and the Minch ice sheets[29].

It has also been suggested that the marine beds of northern Lewis could have been deposited by the Hebridean ice-sheet, which could have been diverted seawards by the higher ground in northern Lewis, to return to the land on the lower ground further north[80].

Torridonian, Cambro-Ordovician and other erratics have been found on North Rona, Sula Sgeir, the Flannan Isles, and even St Kilda. This indicates that at its maximum extent, a 'mainland' ice-sheet covered the entire surface of the Western Isles[67,77,78], though it is possible that the ice did not actually over-run St Kilda[28].

Though the ice-sheet was up to 1500m thick over the mountains around Loch Broom[82], the highest striae or whalebacks in Lewis and Harris occur at about 610m on the Clisham, and the highest in the Uists at about 500m on Beinn Mhor in South Uist. Thus the peaks of the Clisham, Mullach an Langa, Mulla-fo-dheas, Usgnaval Mor, Teilesval, Oreval, Ullaval and the Tirga Mor in North Harris and Beinn Mhor, Ben Corodale and Hecla in South Uist would probably have projected above the surface of the ice-sheet as *nunataks*. The highest summit in the Western Isles known to have been glaciated is Sgaoth Aird (559m)$_7$. Heaval (383m), the highest summit in Barra, was submerged beneath eastward-moving ice$_{50}$.

The presence and distribution of morainic drifts in Harris, southern Lewis and South Uist suggest that the local ice-sheet glaciation was followed by a valley glaciation in these areas.

TIMING

As indicated above, the first indications of ice-rafting activity around Rockall have been dated to 2.4 Ma by the Deep Sea Drilling Project$_{65}$. Each glaciation tends to destroy most of the evidence of the preceding one, and we have positive evidence for only three glacial periods in the Western Isles.

We are fortunate in having a chronological reference point for the last glaciation in the form of an organic layer beneath 'foreign' glacial till at Tolsta Head, which has yielded a radiocarbon date of 27,333 ± 240 years BP (Before Present)[1]. This layer seems to have been deposited in a warm phase during a glacial period (an *interstadial*) rather than a warm period between glaciations (an *interglacial*)$_{83}$.

The most probable sequence of events during the last glaciation is:

a) An ice sheet or ice sheets from the mainland or the Minch covered all the islands of the Western Isles, including St Kilda, possibly on more than one occasion, some time around 27,000 BP.

b) An ice-sheet from the mainland crossed the Minch some time after 27,000 BP, reaching only Tolsta Head and the Ness area, probably at the same time as

c) the Hebridean ice-sheet spread out from three centres in southern Lewis and North Harris, and from an ice-shed lying to the west of the present watershed in the Uists and Barra, from approximately 18,000 to 14,000 BP. This ice-sheet covered the surface of all the main islands except for northern Lewis and Tolsta Head. At this time there was probably a valley glaciation on St Kilda.

d) Local valley glaciation in Harris, South Uist and southern Lewis.

It is difficult to date the earlier ice-sheets, but they are likely to have spanned a

---

[1] For those familiar with the convention of giving callibrated radiocarbon dates in upper case (BP Before Present or BC Before Christ) and uncallibrated in lower case (bp or bc), I have not distinguished between the two, and readers who wish hard and fast information on these dates are referred to my original sources.

period of many thousands of years. Judging by our knowledge of temperatures in the North Atlantic[55], these glaciations probably occurred more than 130,000 years ago. Climatic trends interpreted from deep-sea cores also suggests that valley glaciers are likely to have developed on many occasions on the higher ground, but all traces of these have been destroyed by later glacial activity.

The Hebridean ice-sheet glaciation almost certainly occurred during a period known as the 'Late Devensian', at the same time as the last mainland ice-sheet glaciation, and the coldest part of this glacial phase may have occurred as recently as 18,000 BP[64]. This ice-sheet glaciation lasted from about 27,000 BP till about 14,000 BP[55]. As described above, it has been suggested that the local ice cap and the ice cover of northern Lewis may not have been contemporaneous, so that an area of Lewis could have remained ice-free during the local glaciation, with intense periglacial activity (see below) around the edge of the ice sheet[25]. Based on work done on Trotternish in Skye, Ballantyne[3] has suggested that at the Late Devensian glacial maximum the mainland ice sheet merged with the Outer Hebrides ice cap in what is now the Minch.

The later valley glaciation may be contemporaneous with the Loch Lomond Readvance of the mainland, which is thought to have ended about 10,200 radiocarbon years ago[70], but it is believed that the last Western Isles valley glaciation predates the Loch Lomond Readvance, being correlated instead with the Wester Ross Readvance, which took place about 13,500 BP. The valley glaciation may even have followed the local ice-sheet glaciation without a break[51,52]. These correlations, however, are entirely speculative, as there are no dates for the Outer Hebrides valley glaciation.

PERIGLACIAL FEATURES

In the cold conditions around the margins of a glacier, or during cold winters following glacial retreat or just prior to an advance, or simply during colder, non-glacial episodes of the Quaternary, frost damage to rocks and soils can create characteristic *periglacial* deposits.

Fossil periglacial activity, in the form of frost-shattered stones and deposits distorted and moved by frost action (*cryoturbation*) occur in the raised beach deposits of northern Lewis[81,52].

Frost-shattered debris may accumulate on slopes as *boulder lobes* or at the base of slopes as *scree*[66]. Such lobes, up to 3m in height, occur in the Western Isles above altitudes of about 400m, on slopes between 10° and 25°, and examples may be seen on Gillival Glas (NB149022), Beinn Dubh (NB090007), Loch nan Eang (NB145082) and other peaks in North Harris; on Gormol, Pairc (NB301069); on the south slopes of Mealisval, Uig (NB023270); and on Hecla (Fig. 3.11) and Ben Scalavat in South Uist[52].

Stewart Angus, May 1985

Figure 3.11. Boulder lobes on Hecla, South Uist

Stewart Angus, March 1991

Figure 3.12. Solifluction terraces, Roineabhal plateau

Soil creep in frozen conditions may give rise to *solifluction terraces*, examples of which may be seen on the north-east slopes of Teilesval (NB128099), on the north-west slopes of nearby Usgnaval Mor (NB119088), on the north-east ridge of Oreval (NB085111, 087112), and near the summit of Roineabhal (NG047857)$_{52}$(Fig .3.12).

CHANGES IN SEA LEVEL

Though only 10% of the Earth's land surface is presently covered by ice$_{54}$, sea level would rise by 40-60m if this melted$_{86}$, and expansion in volume when the temperature of the sea water exceeds 4°C would make a major contribution to any sea level rise. At the maximum extent of the Quaternary glaciation, about 31% of the Earth's land surface was ice-covered, so that a correspondingly larger amount of the Earth's water would then have existed as ice, so that sea level would have been much lower than it is today. An extensive submarine platform around St Kilda suggests that sea level may have been about 120m lower than it is today during the Late Devensian (Hebridean ice-sheet, St Kilda valley glacier) glacial$_{70,74,75}$. The huge amounts of ice lying on top of the land weighed so much that the land surface below was depressed. It has been estimated that the amount of depression is approximately one-third of the maximum thickness of the ice$_{86}$ so that the surface of the Western Isles would have been depressed by up to 200m, while on the mainland depression may have been in the region of 500m. More recent work on the St Kilda Basin (south of St Kilda) has suggested that the sea level there may have been as much as 90-100m below its present level around 10,000 years ago, but it has to be said that this is at variance with other estimates from the Atlantic and North Sea, which would put it at only about half this level$_{53}$. The lower estimate appears to be supported by subsequent geophysical modelling$_{39}$. Such sea level depression has given rise to speculation about the possible colonisation of the Outer Hebrides by large boreal mammals such as Reindeer around this time, possibly to the north of Lewis which some claim was not so affected by glacial activity, but with our current state of knowledge this can be no more than speculation.

On the retreat of the ice, the water found its way back into the sea relatively quickly, so that sea level rose and new shorelines were created. Though freed of its glacial burden, the land surface took much longer to recover than the sea but, as it rose, the land took with it these postglacial shorelines, raising them tens of metres above sea level. Because the land has moved as well as the sea, geomorphologists refer to 'relative sea level'.

Classic examples of postglacial raised shorelines occur in the Inner Hebrides and on the west coast of the mainland, and several fine examples may be seen from the Stornoway-Ullapool ferry as it travels up Loch Broom.

In north-west Lewis, raised rock platforms, backed by a degraded cliff and occasionally sloping seawards, have been identified at Swainbost (NB506641), South Dell (NB480626), Asmagarry (NB463609) and Melbost Borve (NB413582) at heights of between 8.6m and 10.1m above the modern cliff bases. Similar shorelines occur at Lower Bayble (NB527304) and Sheshader (NB558338) in the Eye Peninsula, at heights of 6.8m and 6.3m above the modern cliff base respectively [81].

'Raised beach' gravels occur at many localities along the north-west coast of Lewis, as far south as Ballantrushal[81]. These often lie on top of raised rock platforms, and must therefore have been deposited after the platforms were formed[52]. Some of these raised gravels lie on top of glacial till containing Torridonian sandstone, quartzite and Moine erratics, as at Cladach na Luinge (NB 46613) and Toa Dibidale, while others lie on frost-shattered rock ('head'), as at Melbost Borve[81,52]. The gravels in the Habost raised beach have been cemented by calcite, so that they form a conglomerate[81]. I have seen similar conglomerate on North Rona which seems to be of Quaternary age. Other raised gravels occur at Sheshader and Bayble in the Eye Peninsula[22,81] and at Cliad, Barra (NF673048)[50]. Most of the raised gravels are overlain by glacial till, so they must predate at least one phase of glaciation and cannot be of postglacial age[81,52]; they were probably formed after the 'mainland' ice-sheet had retreated from our islands, but before the local ice-sheet glaciation. Others clearly demonstrate periglacial features such as cryoturbation at their upper levels, establishing that these raised beaches may just pre-date a period of periglacial rather than glacial activity. Raised gravel levels are given in the table below.

Locality	Grid reference	Highest raised beach	Highest modern shingle	Difference
Habost	NB508642	11.1m O.D.	3.9m	7.2m
Swainbost	NB508641	13.7m O.D.	6.4m	7.3m
Toa Dibidale	NB648616	12.3m O.D.	4.9m	7.4m
Toa Galson	NB452603	20.0m O.D.	5.1m	14.9m
Melbost Borve	NB413582	9.7m O.D.	3.5m	6.2m

Table 3.1. Highest raised and modern beach gravel (from von Weymarn 1974, p.98). O.D. = Ordnance Datum, from which heights on Ordnance Survey maps are measured.

The exceptional height of the Galson 'raised beach' gravel may be connected with the fact that it lies at the head of a geo which faces the prevailing south-westerly wind[81] though, on the other hand, the small size of the pebbles suggests that the beach may have been formed in calm conditions associated with winter pack-ice offshore[51]. An Iron Age midden is associated with this raised beach, and the site was made famous by excavations carried out in the 1930s[2].

Though high shingle ridges on the west coast of the Uists may appear to be raised beaches, they are thought to be of much more recent origin[56]. The same is probably true of similar high ridges at Arnol, Loch Ereray and Barvas[81]. Loch Ardvule, South Uist, is impounded by shingle ridges 10–70m wide[7] but the loch was partly drained this century when the sea broke through the northern ridge[35]. The large shingle ridge at Stoneybridge, South Uist, has been breached at least twice in the last sixty years[7].

Further evidence of raised shorelines occurs in the form of abandoned cone-shaped stacks between North Galson and Aird Dell. The largest of these, Toa Galson, is over 30m in height. The rocky knoll in the estuary of the South Galson river is probably an abandoned stack[81].

As most of these 'raised beaches' are covered by glacial till, it must be assumed that they predate the late Devensian glaciation[81,16]. The sequence of events is splendidly summarised by my colleague Dr John Gordon in the Geological Conservation Review volume on the *Quaternary of Scotland*[25] (starting with youngest):

6. Late Devensian features, including till limits, shelly multiple drift sequences and ice-free areas with associated drift deposits.

5. Raised beach deposits (pre-Late Devensian).

4. Periglacial deposits.

3. Organic deposits, possibly interglacial (predating the raised beach deposits).

2. Till (pre-dating the raised beach deposits).

1. Raised shore platform of pre-Devensian age (earliest known Pleistocene feature in Outer Hebrides.

The 16km of coastline between Cunndal (NB512655) and Cladach Lag na Greine (NB387557) is of the greatest importance for its multiple drift deposits which are critical to the interpretation and reconstruction of the late Devensian ice sheet, and as such the site has been listed in the *Geological Conservation Review*[24].

Glaciation

Sea level fluctuations associated with glaciation and melting were obviously fairly uniform over northern Scotland – they could hardly be otherwise – but the land surface of the Western Isles had not been depressed so much by ice as the mainland so that subsequent recovery and uplift were correspondingly lower in the outer islands. Indeed, so minimal was this recovery that it could not keep pace with even the later, slower, rise in sea level, and there are no postglacial raised shorelines in the Outer Hebrides. All the indications are that sea level has never been higher than at present during postglacial times, and that a rise in sea level relative to the land has taken place since Man settled in the islands: there are numerous examples of archaeological monuments which are now being eroded by the sea[56,14]. It has even been suggested that the Western Isles may be sinking: whereas most other areas have experienced a relative rise in land level in postglacial times, the Western Isles have experienced a relative rise in sea level and are now sinking more quickly than other areas[1] (see Chapter 4). Having said this, a possible postglacial raised beach has been described from the south shore of Loch Maddy by Alain Godard[22], in the form of a cliff cut in the moraine and a line of raised shingle 0.50m above the present High Water Mark. Professor William Ritchie[57] has indicated that there is ample evidence elsewhere in Britain for oscillating sea levels, and that such a raised shoreline is by no means incompatible with an overall pattern of submergence, and that such oscillations may even have contributed to the accumulation of coastal sand deposits.

Stewart Angus, July 1983

Figure 3.13. Intertidal peat, Calanais

Submerged or intertidal organic deposits which could only have accumulated above High Water Mark have been identified at more than twenty locations in the Western Isles[e.g.36,37,56,81,59,61]. Most of these organic deposits consist mainly of peat, often with the remains of grass, sedge or reed stems, and twigs – or even branches – of trees. At Bagh na Craobhag (Bay of Small Trees) on the tidal island of Vallay (NF773768) and at Holm, Stornoway (NB462317), the intertidal peat contains numerous birch stumps which appear to be *in situ*[6,81].

The spectacular deposit of intertidal peat at Calanais (NB217340) (Fig. 3.13) consists mainly of peat, but also contains fragments of the moss *Drepanocladus revolvens*, a species which is usually associated with nutrient-rich fens[81]. The organic layers at Borve, Benbecula (NF765498) and Peninerine, South Uist (NF732347) contain twigs, and fragments of fresh water mollusc shells[56].

Cemented sand was first noted by William MacGillivray[45] and has subsequently been found on the beach at Luskentyre (NG068989), Borve and Scarista[62]. Though the precise way in which this deposit was formed is not fully understood[63], it is unlikely to have developed below High Water Mark (see also Chapter 8).

Several of the organic deposits have been levelled and dated, and the results are summarised in the table below.

Locality & Source	Grid Reference	Level	Age
Holm, Stornoway von Weymarn 1974[81]	NB46273170	-2.95 to -3.06m	8802 ±70
Forthvath Reef, Pabbay Ritchie 1980[60]	NB909876	"2m below seabed"	8330 ±65
Borve, Benbecula Ritchie 1966[56]	NB765498	-2.0 O.D.	5700 ±170

Table 3.2. Levels and ages of submerged and intertidal organic deposits.

Analysis of the base of the Holm deposit suggests that peat formation began there about 8800 years ago, and that sea level was then at least 5m lower than it is now[81]. Though it is important to emphasise that the Borve (Benbecula) results are not *directly* comparable with those from Holm, the results suggest that sea level could have risen to within 2m of its present level by 5700 BP[56].

The Western Isles could not have been connected to the mainland or Skye by a land bridge during postglacial times. Even the shallowest route across the Minch involves depths of up to 100m, and conventional wisdom is that postglacial sea level depression did not even approach this level[33] (Fig. 314).

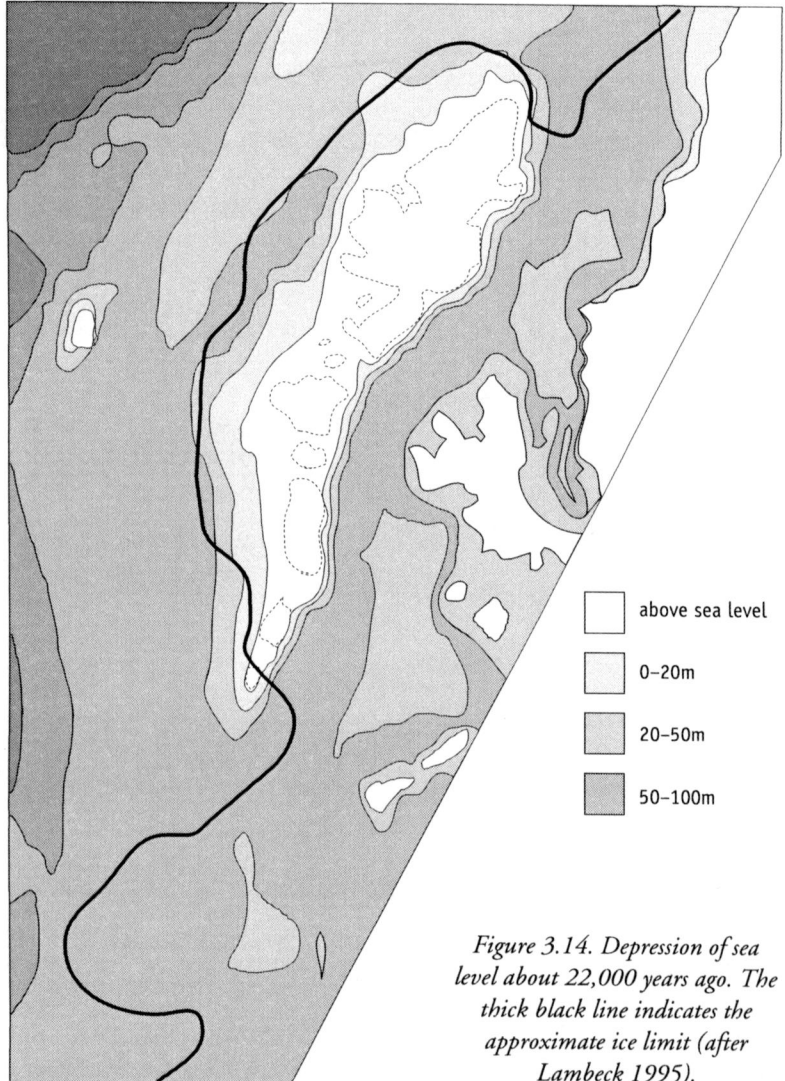

Figure 3.14. Depression of sea level about 22,000 years ago. The thick black line indicates the approximate ice limit (after Lambeck 1995).

above sea level

0–20m

20–50m

50–100m

On the basis of the available information it is possible only to speculate on the possible timing of the final separation of the major island groups of the Outer Hebrides.

Martin Martin[47] tells of a legendary female warrior who lived on St Kilda at a time when "all the space betwixt this isle and that of Harries [Harris] was one continued tract of land". Regrettably, this colourful story cannot be true, for the Amazon's hunting ground lies even deeper beneath the waves than the bed of the Minch.

There is abundant anecdotal evidence of a relative rise in sea level on the western seaboard of North Uist, where submerged features of undoubted terrestrial origin are frequent, and it has even been suggested that the final separation of the Monach Isles from North Uist may have occurred within historic times[1] (see also Chapter 5). The seabed between these islands and North Uist has indeed the form of a submerged isthmus, but most of its area lies at a depth (in relation to Ordnance Datum) of 4-7m, with a 9m channel, which suggests that the sea breached this isthmus more than 9000 years ago, before Man arrived in the Western Isles, but the channel could have been deepened by currents, allowing for a more recent separation. (Note: Admiralty Chart Datum lies approximately 2m below Ordnance Datum.) Further loss of land to the sea is covered in Chapters 5 and 6.

Much of the Sound of Harris lies less than 5m below Ordnance Datum, with a channel 7m deep. This suggests that the Uists were separated from Harris more than 9000 years ago, but more recently than the Sound of Monach was breached. The Sound of Barra is similar to the Sound of Harris in depth, and is likely to have been flooded at about the same time.

Changing sea levels have had a profound effect on coastal morphology. A rise in sea level is thought to be responsible for the formation of many saltmarshes[62,58], and it has been said that the sea lochs of North Uist would be more accurately described by the term *fjards*[57]. Unlike fjords, fjards are surrounded by low-lying countryside and have sinuous coastlines and islands. This is covered more fully in the next Chapter.

OFFSHORE GLACIAL DEPOSITS

It is now believed that at its maximum extent at least parts of the Late Devensian ice-sheet extended westwards as far as the shelf edge and even on to the Wyville-Thomson Ridge, so that the type of sediments associated with ice margins were deposited on the continental slope[72]. Flows deposited particularly thick beds as huge sediment fans on the shelf edge west of Barra and west of Sula Sgeir, up to 600m and 300m thick respectively. Generally speaking, the older beds lie out towards the shelf slope, and inland they are overlain by a sequence of progressively younger deposits. The oldest beds are the Macleod sequence, up to Middle Devensian in age, which are on the slope, replaced by the Conchar sequence on the shelf edge, or the Macdonald beds to the north. Further inland, these surnamed beds are overlain or replaced by beds with male forenames such as Conan and the Fionn of Late Devensian (deposited on the retreat of ice from the St Kilda basin) and Flandrian (after valley glaciation) age respectively. Nearer the coast, in the North Lewis basin, immediately north of the Butt, female forenames take over, and we have (in order of decreasing age) Moira, Shona, Flora, Elspeth, Ailsa, Jean and Morag, but it is well beyond the scope of this book to describe these layers in any detail. Interestingly, much of the shelf to the west of the islands, the shelf to the south of the latitude of Griminish Point in North Uist, is predominantly bedrock$_{72}$ (Fig. 3.15).

The Outer Hebrides ice-sheet of the Late Devensian probably did not reach the shelf edge, and its western limit is marked to the south of St Kilda by a large bank of moraines associated with the Conan beds. The St Kilda basin is infilled by beds of the Caoilte sequence. The St Kilda Basin and the Flannan Trough were formed by glacial overdeepening during the Late Devensian ice-sheet glaciation$_{72}$.

In the Minch, the deposits are divided by the Rubha Reidh Ridge into the North Minch Basin and the Sea of the Hebrides. The Outer Hebrides ice-sheet is believed to have extended north to the latitude of Tiumpan Head in the Minch, with the associated moraines deposited over a wide area of the North Minch Basin as the Fiona sequence, or Greenstone Ridge. To the north of this lie the Morag and Sheena beds, the latter representing deposition in a marine bay to the north of the ice-sheet.

South of the Rubha Reidh Ridge the Stanton beds appear to have been eroded by glacial action, though the Barra formation, which is up to 130m thick, is thought to have been deposited after the late Devensian maximum$_{18}$.

GLACIATION, LANDSCAPE AND MAN

The action of successive glaciations has rendered the already inhospitable landscape of impervious, acid rocks more inhospitable still by the complete removal of sediments from some areas and by depositing extensive beds of till in others; not only is most of the till nutrient-deficient but it has encouraged the widespread formation of peat (see Chapter 8). Glacial roughening has created large areas where only low-intensity land uses are possible; the relatively level, low-lying coastal platforms and northern plateau are difficult to drain, and coastal erosion due to relative sea level changes is often a problem.

The legacy is not completely negative: peat has its uses and the tills of the Eye Peninsula and Ness are more fertile, due to the presence of sediments derived from the Stornoway Formation and shelly beds respectively (see Chapter 8). The sediments washed ashore have created the machair, without which the history of Man in the Western Isles would have been very different.

The till of the Eye Peninsula was exploited commercially in the nineteenth century by Sir James Matheson for the manufacture of bricks and

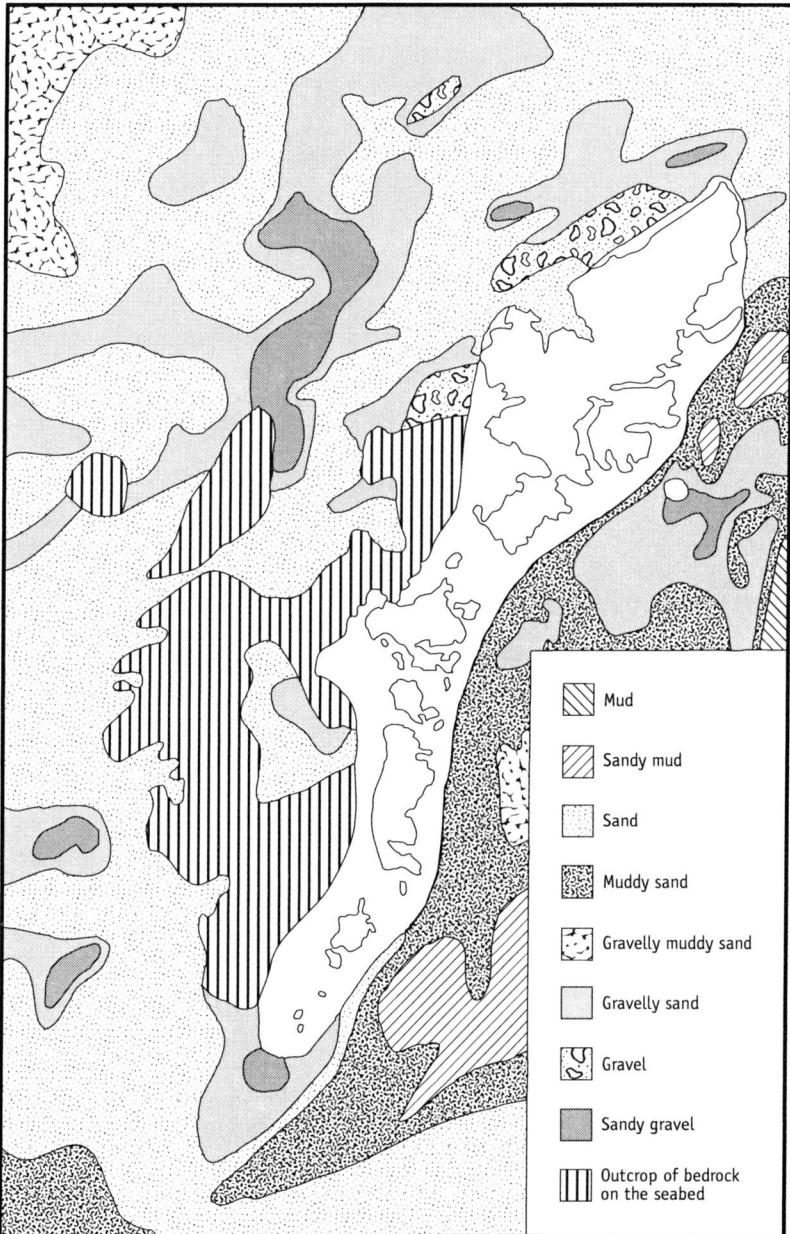

Figure 3.15. Offshore sediment map (after British Geological Survey 1987. Sea bed sediments around the United Kingdom [1:1,000,000, North and South Sheets]. British Geological Survey, Keyworth). Reproduced by permission of the British Geological Survey.

Legend:
- Mud
- Sandy mud
- Sand
- Muddy sand
- Gravelly muddy sand
- Gravelly sand
- Gravel
- Sandy gravel
- Outcrop of bedrock on the seabed

drainage tiles. The clay pit and brickworks can still be seen at Garrabost (NB 508331)[32].

The main source of shingle for building was until recently the spit at Teanga Tunga (Tong) but extraction has now completely ceased there, and almost all the shingle required in Lewis is taken from Carnish in Uig (Fig. 3.16). Building sand has also traditionally been collected from the beaches around Stornoway, but the scale of extraction was such that the Local Authority leased the foreshore from the Crown Estate Commissioners, and imposed Coast Protection Orders on most of the beaches in Broad Bay, so that since 1970 a permit has been required for the extraction of sand below High Water Mark[62]. There is also a Coastal Protection Order on the beach adjacent to Benbecula Airport[12].

Stewart Angus, April 1992

Figure 3.16. Carnish gravel quarry

Sand is extracted on a large scale at Eoropie, Barvas, Horgabost, Balelone, the North Ford, and Tangusdale. Planning permission is now required for all commercial mineral extraction but the removal of exposed sand or gravel by crofters for their own use does not need the consent of the planning authority, though in most circumstances the permission of the proprietor is required. Many grazings committees are now aware of the potential dangers of *ad hoc* mineral extraction, and have restricted or even banned the removal of sand. Machair sand, by virtue of its high shell content, is particularly suitable for neutralising the acid moorland soils, and has been extensively used within the last few decades in moorland reseeding schemes. Reclamation of peatlands is no longer encouraged, and such extraction has virtually ceased. The potential threat of unregulated sand and gravel extraction is now widely recognised, and as part of its preparation for a Minerals Strategy for the Western Isles, Comhairle nan Eilean commissioned a report on the sand and gravel resources of the Outer Hebrides from the British Geological Survey, assisted by Scottish Natural Heritage and Western Isles Enterprise. The resulting report[12] gives guidance for the future exploitation of the vulnerable sand and gravel deposits of the Western Isles.

The rate of extraction of gravel from Carnish had begun to worry conservationists by the 1970s, and is said to have increased dramatically during the extension of the main runway at Stornoway Airport in the early 1980s. The Carnish Terrace is afforded a certain degree of protection by virtue of the fact that it partially impounds Loch Scaslavat, but a proposal (later dropped) by Redland Aggregates Ltd in 1992 to develop a large coastal quarry to exploit this deposit allowed a thickness of only 6m of sediment as protection for the loch.

The possibility of oil-related development in the Western Isles was partly responsible for the production of *Outer Hebrides: Localities of Geological and Geomorphological Importance* by the Nature Conservancy Council[7]. Not only are geomorphological Sites of Special Scientific Interest fully discussed, but sites of subsidiary interest are also highlighted, so that the islands are unusually fortunate in possessing a comprehensive prospectus for the conservation of geomorphological features.

Neither oil-related development nor lines of wave energy converters offshore seem as imminent as they did a few years ago, but the fragile nature of many of our best physical features requires that even the smallest development affecting them is carefully planned. Detailed study of the NCC report[7], the Countryside Commission for Scotland's valuable beach reports[62,58] and the British Geological Survey report[12] is essential if human, as well as nature conservation, interests are to be protected for the future.

The Western Isles Council's Minerals Strategy gives a high priority to the natural heritage and incorporates much of the above, including the all-important recommendations of the BGS report.

4

COASTAL DEVELOPMENT

After leaving the cove we happened on another of nature's workshops. A peculiar sound as of steam escaping under pressure had been heard at regular intervals, which we found proceeded from a small crack on the face of the cliff about 12 feet above the boat. Lying to for a little, we saw that the falling swell exposed the mouth of a submerged cave which filled with air, to be expelled by the rising wave with the force of an explosion at the minute fissure above our heads, showing that even the adamantine gneiss of the Hebrides is no proof against the persistent attacks of the Atlantic. (Wedderspoon 1912)[54].

From the Appendix it can be seen that the coastline of the Western Isles may be anything from 1813km to 3395.27km in length, a range which demonstrates the difficulty of measuring the intricacies of every inlet. This Chapter accepts the JNCC estimate of 2102km; the different coastal types are shown in Table 9.2. It will be more than obvious to anyone who knows the islands that the coast is forever changing, as some features alter and others disappear as they lose their battle with the might of the sea. The rate of change varies considerably, from very slow in the case of the hardest rocks, to the sediments of exposed coasts, which may undergo considerable transformations over the period of a single storm.

Jost von Weymarn, the man who did so much to throw light on the history of glaciation in the Outer Hebrides (see previous Chapter), also did much valuable work on the long term coastal development of Lewis and Harris, while Professor William Ritchie studied every beach system on the main islands and made detailed examinations of the contemporary coastal development of the Uists. This Chapter owes much to these two fine geographers.

The form of the coastline is related to a range of variables, all of which change over time: 1) geology and structure, 2) the inherited pre-Pleistocene landform, 3) climate, 4) marine variables, and 5) the relative position of sea level, 6) availability of mineral sediment and 7) availability of organic sediment (shell)[41-47,53]. Because these variables operate in conjunction with each other it is not possible to embark on a sequence of descriptions in this order; it is easiest to start off from the end of the last chapter, where the role of glaciation is still of great importance, and even the pre-existing landscape must be recalled at times.

THE INHERITED PAST

Though the entire landscape of the main islands of the Outer Hebrides has been glaciated, there is enough evidence to draw certain conclusions about the derivation of some of the older coastal features. Von Weymarn stated that the sea lochs of central Lewis – Erisort, Roag, Grimshader, and Leurbost – were pre-existing depressions developed along lines of geological weakness, and had not been overdeepened by the ice to form fjords. Broad Bay and the Minch itself owe their existence to *grabens*, where a whole section of rock has been faulted downwards in relation to the surrounding rocks (see Box 2.5).

Most of the rocks of the Outer Hebrides are extremely durable and do not readily break down. Erosion which has taken place since the retreat of the ice has acted upon a pre-existing coastal landscape, with ready-made lochs, valleys and cliffs, with the complication of different sea levels at different times. Foliation planes and lines of jointing and crushing have been exploited, so that the generally straight north-west coast of Lewis follows structural trends, as does the Sound of Harris coastline of Harris which is obviously aligned in the same direction as the metasediments[53].

On the shore, vertically or steeply dipping rocks tend to promote the formation of stacks and successive reefs, as in parts of north-west Lewis.

Igneous dykes are often lines of weakness and many caves, geos and tunnels may be related to preferential erosion of these local structures. The BGS 1:100,000 geological maps show that the great majority of Tertiary igneous intrusions intersect the coast at right angles, and are thus exposed to marine erosion.

A GLACIATED COAST

The impact of the ice on the coastline varies a great deal. Some of the sea lochs seem to echo the wide inland valleys of North Harris and Uig, and indeed these lochs are their marine equivalents, for the fjords are no more than submerged glacial troughs (see Chapter 3). Jost von Weymarn has pointed out that the term *fjord* is usually applied to features on a greater scale than the sea lochs of the Western Isles, but the term is now so universal that its use can be justified here. The key feature of a fjord is that it has a deep basin, but shallows towards the sea to a *sill* which may be very close to the surface. Lochs Claidh, Seaforth, Shell and Tamanavay are fjords. Separation of Scalpay and the isolation of Seaforth Island

Stewart Angus, September 1977

Figure 4.1. The 'drowned' fjardic coastline of Loch Maddy, North Uist

are the result of glacial overdeepening; with a maximum depth of 98m, Loch
Seaforth has the deepest basin within any sea loch in the Outer Hebrides. Loch
Eport is steep-sided and narrow, and it does not follow structural trends; it is
30m deep at its entrance and has no sill, and it is likely that its origin is related
to faulting followed by ice movement[43]; the absence of a sill may be related to
the presence of the deep Minch channel immediately off the mouth. Gregory[22,23]
believed that Loch Eport represented an asymmetric valley, with a scarp caused
by a geological fault on one side and the drowned countryside associated with
a *fjard* on the other. Unlike fjords, fjards are surrounded by low-lying countryside and have sinuous, 'drowned' coastlines and islands. Loch Maddy (Fig. 4.1)
is one of the best examples in Europe of a fjardic coastline, and the tidal range
of 4m, combined with the great variety of 'inland' lochs on the sinuous coastline,
has created an almost unique assemblage of lochs of varying salinity. In nature
conservation terms, it would be hard to understate the importance of the Loch
Maddy fjard, and much of the biological interest is related to the physical
structure of the coastline. Loch Skipport has been described as a fjord with
fjardic features[26].

EROSION OF HARD COASTS BY THE SEA

Breaking waves may exert a pressure of 8 tonnes/m² and in severe winters the pressure may exceed 24 tonnes/m²[25]. Off the west coast, wave heights exceed 3m for 10% of the year[15]. Spray and air are driven into every crevice under this immense pressure, expanding explosively as the wave recedes. Any line of weakness – along a fault or intrusion – is exploited. A storm in January 1836 is reputed to have moved a boulder weighing about 42 tonnes a distance of nearly 2m at Barra Head[18].

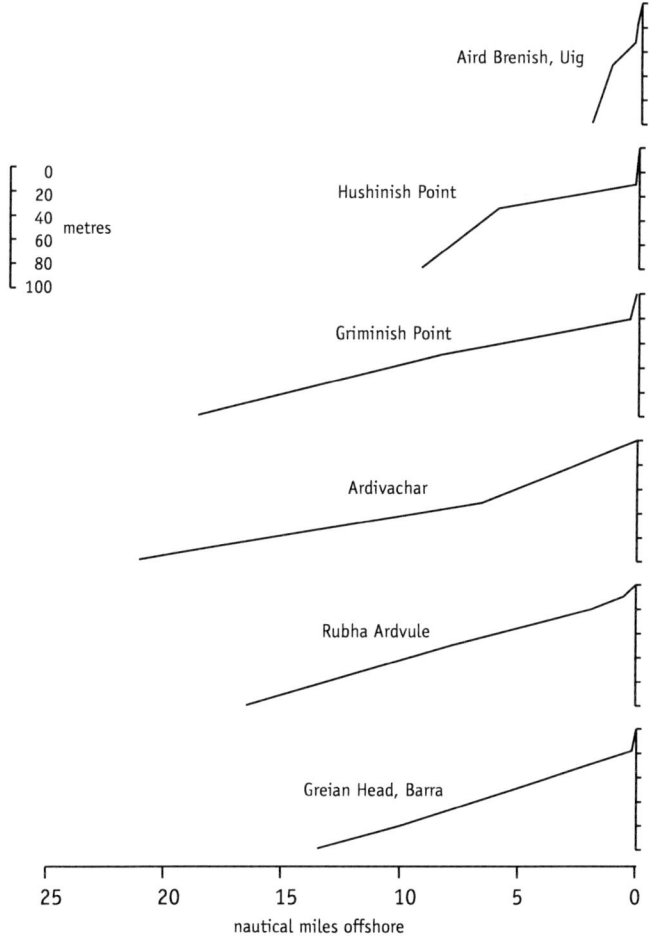

Figure 4.2. Offshore seabed profiles on the west coast of the Outer Hebrides, using a bearing of due west (270° true)

Seabed profile (Fig. 4.2) may be very important in determining the impact of wave action: the cliffs of Uig and St Kilda lie adjacent to comparatively deep water, and are thus exposed to the full force of Atlantic breakers, whereas the seabed west of South Uist is gently sloping and supports a vast forest of tangle or kelp *Laminaria hyperborea*. This shallow seabed and the seaweed which flourishes upon it combine to absorb a great deal of wave energy before it reaches the west-facing shorelines of the Uists.

Some beds of the Stornoway Formation are stronger than others, extending out to sea as platform-like reefs while the softer material has yielded more easily to battering by the seas. Likewise some softer beds on the upper cliffs of the Stornoway Formation may give a convex cliff profile, in contrast to the more usual vertical cliffs of the gneisses (Fig. 4.3). Marine erosion may undercut harder cliffs, and the rocks of St Kilda seem especially prone to undercutting: the 1:10,000 Ordnance Survey map uses the word 'overhang' 35 times on the coast of Hirta alone, giving rise to the intriguing (but unlikely) possibility that the island has a greater area at High Water Mark than Low Water Mark..

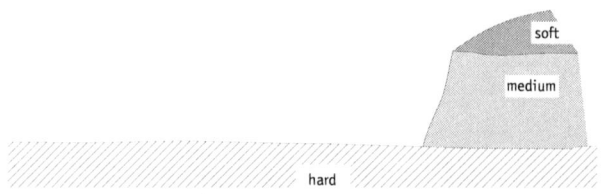

Figure 4.3. Preferential marine erosion of different beds of the Stornoway Formation. A more durable layer often crops out on the shore in Broad Bay, forming a wide platform.

Progressive, localised undercutting (usually of structural weaknesses or softer rock types) forms caves. The development of the cave depends on the alignment of the weaknes to the sea; frequently lines of weakness lie at right angles to the sea (Fig. 4.4). As the cave is enlarged, the front of the roof collapses, the material is washed away, and the process continues, forming a *geo* or *sloc*. The narrow inlets thus formed are very characteristic of the more exposed coasts of the Outer Hebrides, and 'geo' occurs as the first word in 489 named features on the 1:10,000 sheets of the Outer Hebrides, and in many other names as a subsequent word. On Taransay, they are called 'gya', this being the equivalent Norse suffix..

Sometimes, instead of (or as well as) collapsing from the seaward end, the cave roof gives way at the inland end, forming a shaft linking the cave to the

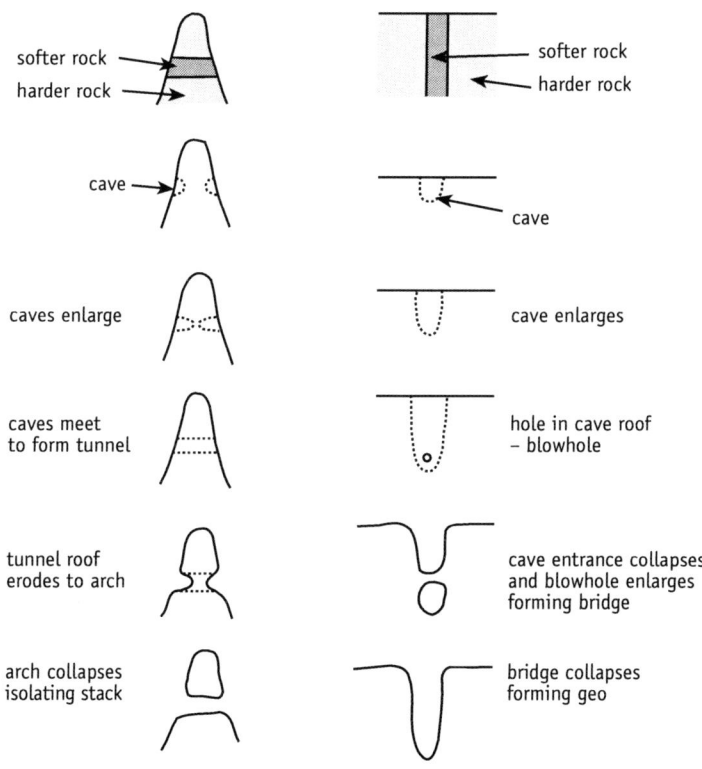

Figure 4.4. Development of stacks and geos

surface. During storms, waves are driven explosively upwards through the 'blow-hole' or *gloup*. A spectacular blow-hole may be seen at Seilaro, south-west of Aird Uig NB 034370. Other examples occur at Dun Eistean, Port of Ness NB 536650[27], at South Shawbost NB 243476 and at Sgeildige on North Rona (Fig. 4.5), where the blow-hole is said to have been "burst out from below by the sea" during a storm in December 1862[50]. William Jolly, who wrote of Mingulay as *The Nearer St Kilda*[29], claimed that local people referred to caves and natural arches as *cailleach* (old woman): "When the sea rushes into one of them, it causes a concussion of air and water which results in a deep hollow snort and upward rush of spray, which is picturesquely described by the natives as 'the old woman taking snuff'". On the north side of Tigharry Point, not marked on the 1:10,000 maps, there is a blow-hole called *Sloc a'choire* (pit of the cauldron), where a water

Stewart Angus, June 1986

Figure 4.5. Gloup, North Rona

spout of up to 200 feet may be seen in westerly storms. There is a local tradition that the bridge forming the arch on the seaward side of the blow-hole will one day give way beneath the weight of a newly married couple[7].

Where caves are excavated on opposite sides of a headland, they may eventually meet, forming a natural arch or, if the headland is wide, a tunnel. (Fig. 4.4). There are just over a hundred natural arches in the Outer Hebrides, of which over 80% are on the more exposed coasts of Lewis and Harris, with a mere seven in the Southern Isles, and six in the St Kilda group. The greatest numbers of arches is on the Uig coast south of Valtos (20) and in the area around the Butt of Lewis (18).

The longest tunnel is through Rubha na Beirghe NB235474, south of Shawbost. At 140m, it is 40m longer than the famous Tunnel on Hirta (Fig. 4.6). One of the great rituals of going to St Kilda is walking the length of the Tunnel, hanging from a securely fastened rope for some of the traverse, under the inquisitive eyes of seals in the surf. I have been out to the tunnel which extends 120m through the island of Campay in West Loch Roag, but it is a bit too narrow to contemplate sailing through, though canoeists may have succeeded. St Kilda boasts a fine double tunnel through the south end of Dun, but perhaps the finest double tunnel is that at Sloc Roe in North Uist, where there is a blowhole 20m in diameter in the middle of a tunnel through Griminish Point. St Kilda also has a number of underwater tunnels, including a 60m tunnel through Levenish, a 40m example in Am Plaistair, and an arch about 30m long

Stewart Angus, August 1993

Figure 4.6. The Tunnel, Gob na h-Airde, St Kilda

through Sgarbhstac. The last has been described as the "world's best dive site"[40].

Some natural arches do not go through headlands or islands but form a bridge across a geo. These really represent a greater development of the blowhole, being a transition form preceding the total collapse which will form a geo (Fig. 4.4). Examples may be seen at Crowlista in Uig NB028340 and at Roagh on the west side of Taransay NB001007 (Fig. 4.7). There is a small geo with a correspondingly small bridge on the island of Boreray in the Sound of Harris NF850804, the bridge of which looks as if it will not be here for very much longer. In his book on Taransay, Bill Lawson points out that structures such as the Roagh arches may be spectacular in the eyes of visitors, but present serious problems for livestock and those looking after them[30].

While the collapse of the cave roof or the bridge forms a geo, the collapse of a tunnel isolates a stack. The infamous Alasdair Alpin Macgregor, in his *Last Voyage to St Kilda*[31], relates a story about an arch which was said to have connected Mina Stac to Conachair until it was struck by the tip of the topsail mast of a ship of the Spanish Armada driven before a gale. The ship was buried in the rock fall. Stacks are indeed formed by the collapse of natural arches through narrow headlands, but we have no way of knowing if Mina Stac owes its premature existence to a Spanish shipwreck, and dives on the site have failed to reveal any signs of a wreck[40].

An excellent series of stacks may be seen at Garry Beach, where one of the stacks resembles Queen Victoria when viewed from the south cliffs (Fig 4.8).

Stewart Angus, July 1989

Figure 4.7. Natural arch, Roagh, Taransay

Stewart Angus, June 1983

Figure 4.8. Sea stacks, Garry Beach, Lewis. Note the profile of the stack on the left, which is said to resemble Queen Victoria.

Stewart Angus, June 1987

Figure 4.9. Stac an Armin, St Kilda, the highest sea stack in the UK

Stewart Angus, June 1978

Figure 4.10. Mangersta cliffs, Uig, Lewis – on a calm day!

The classic site for stacks in the Outer Hebrides, if not Britain, must be St Kilda, where Stac an Armin (Fig. 4.9), at 196.3m, is the highest sea stack in the UK, with neighbouring Stac Li 171.9m; there are also impressive stacks between An Cambir and Soay.

Of the 2102 km of island coastline, there are 96.5km of vertical cliffs higher than 20m and 36km lower than 20m, a total of 132.5km, most of it in Lewis (Table 9.2) (Fig. 4.10). All the cliff types together make up 214.5km, or 10.2% of the coastline. Biulacraig on the west coast of Mingulay is a stunning 229m and until 1878 was believed to be the highest sea cliff in Scotland This cliff formed not only the crest of the Macneils of Barra, but also their rallying cry "Bulnacraig"[10]. The honour of having the highest vertical sea cliff not only in Scotland but in the UK now goes to Conachair in St Kilda (Fig. 4.11), with a height of 426m, though even this is dwarfed by the highest Irish cliffs at Slieve League in County Donegal, which attain a height of 600m.

The process of change on hard rock coasts generally takes place over thousands of years, and change is usually only noticed when there is a sudden event such as a rock fall or the collapse of a cave roof. Change on soft coasts is much more rapid, so that even gradual changes may be apparent over short periods.

Stewart Angus, June 1987

*Figure 4.11. Conachair cliffs, St Kilda, the tallest vertical sea cliffs in Britain.
Only the top third is visible in the photograph.*

SOFT COAST FEATURES

Saltmarshes are intertidal features associated with low energy environments: the fine silt which makes up much of the alluvial soil in saltings can only settle out in slow-moving waters, or be brought into contact with the pre-existing surface on a falling tide. Most of the saltmarshes of the Western Isles are therefore located at bay-heads and at the head of sea lochs, though the latter and the 'fringing' saltmarshes of parts of the Uist fords tend to have a restricted area. Officially (i.e. from Table 9.2) there are 348ha of saltmarsh in the Western Isles, but estimates vary, and there may be as much as 500ha. The largest site is the Melbost-Tong inlet, with an area of 96ha, with significant expanses in Baleshare, Seilebost, Camas Uig, Bayhead (N.Uist), Northton and Gress.

It is likely that these estuarine saltmarshes originated more than 5,000 years ago, when sea levels were slightly lower in relation to the land surface of the Western Isles and the very fine (clay) fraction of glacial deposits supplied suitable sediments for saltmarsh formation. Nowadays, our rivers have catchments which consist of blanket bog and resistant rocks, neither of which yields much in the way of sediment. The seaward edge of most of these marshes often takes the form of a micro-cliff, exceptionally up to 1.5m high, but usually less than 0.5m, suggesting that they were once rather more extensive than they are now. Certainly they now consist almost entirely of 'upper saltmarsh' vegetation, with very little 'lower saltmarsh' or pioneer vegetation.

In addition to the scarp which often marks the eroding seaward limit of a saltmarsh, there are other features which break the regularity of the flat surface. The most widespread of these is probably the 'pan', a depression or pool in the saltmarsh plain, ranging from a few centimetres in depth (and often drying out) to deep, permanent, steep-sided pools. The larger areas of saltings such as Tong and Luskentyre also have creeks – long, narrow channels which are often sinuous in form. Work in Shetland and Norway suggests that the creeks act primarily as drainage channels for surface water rather than as two-way tidal channels[13]. These features are best seen from above. The road affords a good view of the saltmarsh at Northton (Fig. 4.12) but the best views are undoubtedly obtained from the air (Fig. 4.13), when the complexity of creek and pan is spectacularly displayed.

Detailed studies at Tong and, to a lesser extent at Uig and Luskentyre[52] suggest that creeks originate from large desiccation cracks created during periods of drought, in a process analogous to the formation of peat hags. Pans are thought to have been created by the collapse of natural drainage 'pipes' of sandy material some 40–70cm underground.

Stewart Angus, September 1979

Figure 4.12. Saltmarsh at Northton, Harris

Stewart Angus, August 1987

Figure 4.13. Aerial view of Gress Saltings, Lewis

Marine erosion of the Stornoway Beds and their overlying till on the northern side of Broad Bay has yielded large quantities of sand and gravel which the sea has carried south by longshore drift, whereby obliquely breaking waves transport sediment along the coast. Shingle from these sediments has accumulated in the form of a spit at Teanga Tunga south of Tong. There is little accumulation of coarse sediment on the spit at present, and the spit has been modified by the changing courses of the Abhainn a'Ghlinne Dhuibh and the Laxdale River[48].

The Branahuie isthmus which links Point with the rest of Lewis is thought to have been formed by the erosion of till during postglacial sea level changes. Boulders washed from the till by rising seas were built into ridges by storm waves and later covered by finer sediments[53] (Fig. 4.14).

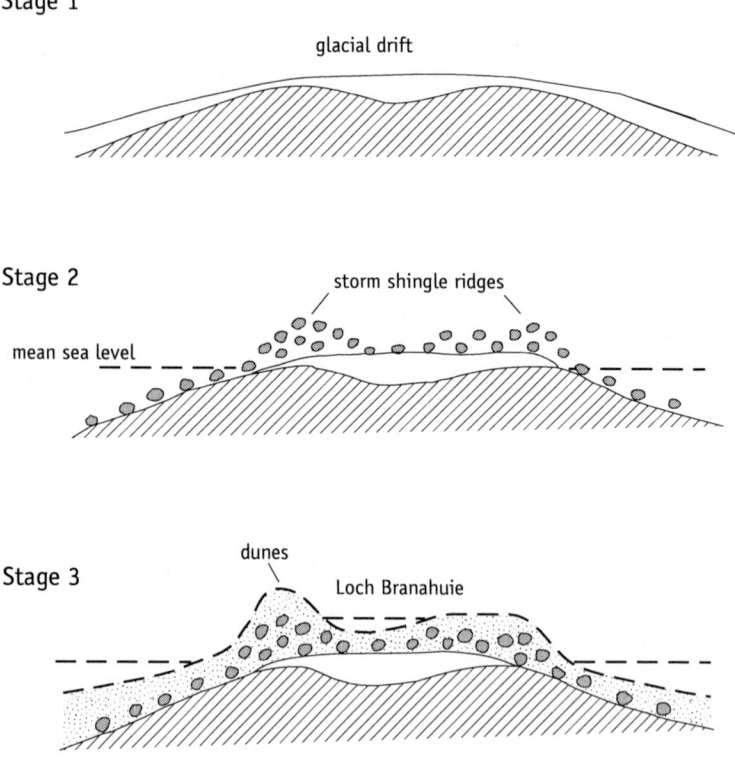

Figure 4.14. Diagram of structure of Branahuie tombolo
(after von Weymarn 1974)

Stewart Angus, October 1984

Figure 4.15. The double tombolo at Holm, Lewis, with lochs at each end. Since this picture was taken the southern loch (left) has been filled in by works associated with the building of the MOD jetty.

Gob Shilldenish, Holm NB 460310 is linked to the mainland by two shingle *tombolos* (Fig. 4.15). As at Branahuie, the ridges impound a small loch$_8$. The Branahuie ridge is believed to be a 'fossil' feature, in that it is now subject to severe erosion, and was obviously constructed when currents or sediment availability were different. The isthmus on Vatersay is a very fine example of a sand-built tombolo$_8$. The Monachs are linked to North Uist underwater by a tombolo-like ridge which may even have been a land link within recent times (see Chapters 3 and 5), and the same may be true of the submerged ridges between Corran Raah on Taransay and Luskentyre on South Harris; between Scarp and Tràigh Mheilein in North Harris; and between Verran Island and South Uist.

Groynes – man-made barriers aligned at right angles to the sea – have been built on both sides of the Branahuie isthmus in an attempt to combat westward longshore drift, with the intention of trapping sediment at the eastern end of the two beaches, but these no longer function adequately (Fig. 4.16). Without the protective sea walls, it is quite conceivable that Point could be cut off from the rest of Lewis by erosion. Langa Sgeir Mhor, on the Broad Bay side of the isthmus, has the effect of a gigantic natural groyne. Erosion of the Braigh isthmus is discussed in Chapter 6.

Stewart Angus

Figure 4.16. Groynes, Branahuie isthmus. Broad Bay is on the left of the picture.

The isthmus at Northton has clearly developed in a quieter environment, possibly linked to saltmarsh formation, deposition of blown sand, and changing water levels. The north side of the isthmus has a series of curved bars, each probably representing a stage in coastal advance; these are most easily seen in winter, when they are often separated from each other by pools.

One of the most unusual features of the Tràigh Mhor in Barra is the presence of a number of cusp-shaped ridges composed entirely of empty cockle shells *Cerastoderma edule*. The position of these bars has changed over the years, and incoming tides in quiet conditions can cause large numbers of upturned shells to float into new positions[17]. Until recently some 600 tonnes of cockle shells were extracted from this beach each year and processed at Suidheachan, the nearby house originally built by Compton Mackenzie. The excavation of this shell material affected drainage on the beach, and had begun to concern the authorities as long ago as 1966[42].

The peninsula of Corran Seilebost in South Harris has the superficial appearance of a spit, but is thought to be a surviving part of a drowned dune and machair complex which once occupied Tràigh Luskentyre[48]. The soils of the Seilebost saltmarsh are characteristic of machair[34] which supports the drowning hypothesis. Gualan, at the northern tip of South Uist, seems to have several 'hooks' incorporated in its structure but, like Corran Seilebost, it is thought to be all that remains of a drowned machair$_8$. The tip, at least, of Slugan, at the north end of Baleshare, appears to be a genuine wave-created spit, though the southern section may be the surviving part of the extensive machair which once occupied the sand flats area to the east of Slugan$_{43,44}$. Hooks are visible at the extremities of both Baleshare and Kirkibost$_{43}$ and two hooks are discernible in the northern section of Teanga Tunga$_{48}$. There is a remnant of machair at the end of Corran Aird a'Mhorain in North Uist (Fig. 6.15), suggesting that it may be the surviving part of a machair which once extended over Tràigh Ear$_{43}$. The word 'Corran' in place-names denotes a sickle, reflecting the hooked shape of these spits; earlier hooks are often discernible on the inland side of the spits, making a series. The 'hook' of the spit is caused by the refraction of waves around the tip (Fig 4.17). The machairs of Gress and Coll may be built on old shingle spits$_{48}$.

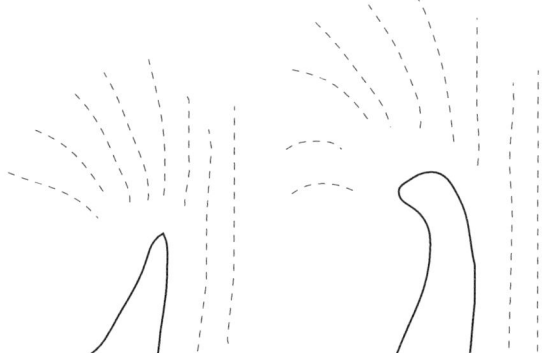

Figure 4.17. Refraction of waves round tip of spit, causing development of 'hooks'

Considering the amount of glacial erosion from the land surface of the Western Isles, there is very little in the way of glacial deposition on the land. Terminal moraines, the last deposits left by glaciers prior to melting, are almost totally absent, while deposits left by glacial meltwater are scarce. The obvious conclusion is that the glaciers kept on moving into a sea that was then rather lower than it is today, and the terminal moraines and gravels left by meltwater were deposited on what is now the seabed.

The Lewisian rocks are hard and unyielding, so that contemporary erosion of rocks contributes little to shore or offshore sediments, though till beds may supply shingle in some areas. Only in and around Broad Bay are there significant sediments derived directly from the surrounding rock, in the form of shingle deposits, as on the south side of the Braigh isthmus, or in the form of the mineral sand which comprises most of the beaches of Broad Bay. This is noticeably darker in colour than the shell sand of the west coast beaches, being derived from marine erosion of the sandstones of the conglomerates of the Stornoway Formation rather than shell fragments and glacial debris.

Other large stones and boulders from glacial deposits have found their way on to shingle and cobble shores (Box 4.1). There is a spectacular sequence of coastal lochs on the west coast of Lewis (Fig. 4.18) formed by the impoundment of streams by cobbles, extending from Loch Mor Barvas south to Loch Dalbeg; the cobble ridge at Dalbeg is covered in sand. Shingle also impounds lochs on Opsay and Boreray in the Sound of Harris. A line or ridge of shingle and cobbles forms a significant part of the upper beach on much of the Atlantic

BOX 4.1 – Sizes of particles	
Particle type	**Diameter (mm)**
Boulder	>256
Cobble	64-256
Pebble	4-64
Granule	2-4
Very coarse sand	1-2
Coarse sand	0.5-1
Medium sand	0.25-0.5
Fine sand	0.0625-0.25
Silt	0.0039-0.0625
Clay	0.00098-0.0039

Stewart Angus, August 1987

Figure 4.18. Loch a'Bhaile, Shawbost, one of a number of lochs on the west coast of Lewis impounded by a shingle ridge

seaboard of the Uists, the most dramatic development of this ridge being at Stoneybridge, where a cobble rampart has been built to a height of 7m. This shingle ridge is of vital importance wherever it occurs, because it absorbs a significant part of the energy from storm surge waves, and forms a strong physical barrier protecting the machair. Even so, the Stoneybridge rampart has been breached by the sea at least twice this century. Many (though not all) of the boulders in the Lewis deposits are angular rather than rounded, demonstrating that they probably do not have an offshore origin, though they have probably been reworked locally by marine activity; the uppermost boulders often support a mature lichen flora, suggesting that they are not moved often by storms. Where they occur, the more rounded boulders are likely to have been shaped by being rolled against each other by wave action. There is a boulder beach on the west side of Fianuis on North Rona which is often described as a storm beach, but these boulders are also angular, so that it is not quite a true storm beach (Fig. 4.19). The line of boulders at Mangersta has been described as a storm beach by Ritchie and Mather[48] who said it was "composed of coarse, sub-angular cobbles derived locally and built into a ridge form across the mouth of the depression". At Garry in NE Lewis, the boulders at the south end of the beach seem to be derived from erosion of the cliff above[48]. The pebbles at Arnol seem to be well rounded and exhibit signs of sorting with a tendency to smaller pebbles below and larger above, so that this is possibly a genuine storm beach.

Stewart Angus, November 1995

Figure 4.19. 'Storm Beach', Fianuis, North Rona

The shingle ridge at Aird in Benbecula has been described as one of the "finest storm shingle accumulations anywhere in Britain"[44,8]. Rounded boulders up to half a metre in diameter are thrown well inland at high tides during the wildest storms.

Smaller pebbles form shingle beaches or a shingle band on sandy beaches. The word 'Mol', meaning 'shingle', occurs in over 100 place-names in the Western Isles. Shingle and cobbles, probably mainly from 'lag' deposits associated with glaciation, where the finer particles have been washed away, often form the core of peninsulas and spits in dune and machair systems.

It would be argued by many, including myself, that the physical feature which makes the Outer Hebrides special is the machair, that stretch of sandy grassland which borders the Atlantic. Machair is of outstanding nature conservation interest for its landforms and its plant and bird life, and makes living in these areas so very much easier than in the parts of the islands where machair does not occur.

It is necessary at this stage to define machair, as it is still not clear, even to many biologists, just how it differs from other coastal grasslands.

The word 'machair' is given a range of meanings by Dwelly, in his classic Gaelic-English Dictionary, from "the low and level part of a farm" to "long ranges of sandy plains fringing the Atlantic side of the Outer Hebrides".

Even ecologists cannot agree on how they use the word, but I suppose we should be grateful that they have not mangled it or even replaced it, and machair is probably the only major habitat to have a Gaelic name. Some purists insist that the term 'machair' applies only to the plain, and does not include the dunes and to avoid confusion, the plain itself is now by convention referred to among scientists as the 'machair grassland'[12]. The machair grassland does not exist in isolation, but is closely linked to the beach, dunes, hill machair, machair lochs, saltmarsh, sand flats and even saline lagoons, blackland (linking machair grassland and moorland) and the transitions between these, and the whole is now referred to as the 'machair system'[2]. The term 'blackland' has been used by some authors to describe moorland, and by others to denote the transitional zone between machair and moorland, often used for hay cropping and where much of the human settlement is concentrated. To resolve this ambiguity, I consulted the resident Uibhisteach in our office, Johanne Ferguson, who advised me that "blackland is a term invented by scientists to describe the area between the machair and the moor". So there.

Professor Ritchie's 'official' definition of machair[46] is given in Box 4.2, and the number of criteria required is an indication of the complexity of the habitat. Even now, there is some disagreement over the precise geographical limits of machair in Scotland – does it, for example, include the beaches of the north coast of Sutherland or even Caithness – does it include Orkney systems

---

BOX 4.2 – Definition of machair (from Ritchie 1975)[46]

1) a base of blown sand which has a significant percentage of shell-derived materials
2) lime-rich soils with pH values normally greater than 7.0.
3) a level or low-angle smooth surface at a mature stage of geomorphological evolution.
4) a sandy grassland type vegetation with long dune grasses and other key dune species having been eliminated. Core plants are Red Fescue *Festuca rubra*, Common Bird's-foot-trefoil *Lotus corniculatus*, White Clover *Trifolium repens*, Yarrow *Achillea millefolium*, Lady's Bedstraw *Galium verum*, Ribwort Plantain *Plantago lanceolata*, Eyebright *Euphrasia officinalis*, Daisy *Bellis perennis* and the moss *Rhytidiadelphus squarrosus*.
5) biotic interference such as is caused by heavy grazing, sporadic cultivation, trampling and sometimes artificial drainage should be a detectable influence within the recent historical period.
6) an oceanic location with a moist, cool climatic regime.

---

such as those of Sanday – or Quendale in Shetland? Machair is listed by the EC Habitats and Species Directive, and the general description from the UK's list of possible Special Areas of Conservation[28] is given in Box 4.3. Though a more detailed description has been prepared by the European Commission[14] this is almost entirely biological.

BOX 4.3 – General description of machair (JNCC 1995)[28]

Gently sloping coastal plains formed by wind-blown calcareous shell-sand, which incorporates a mosaic of species-rich grassland, fens and lochs, with dunes towards the sea and blackland (a mixture of peat and sand) further inland. All these individual elements are crucial parts of the machair system. The grassland has traditionally been maintained by low intensity agriculture. In Europe this habitat occurs only in Scotland and Ireland.

The first and most obvious distinction is the high proportion of shell fragments (Fig. 4.20) in machair sands, not just on the beach, but in the inland soils, though the proportion of shell decreases with distance from the sea and with the age of the soil, and also varies from beach to beach. This is not enough, however: some Hebridean machair systems have comparatively low amounts of shell, while some systems on the Scottish mainland may have high amounts of shell, but are not machair.

The second, and possibly most crucial element, is that of climate. Sand only blows when dry, and there must be strong winds to blow the sand inland from the beaches, but there must also be sufficient moisture to trap the sand, and the winds and the moisture must be at just the right times of year to get the right balance between erosion and deposition to form and maintain machair.

Thirdly, and also of vital importance, is the role played by people. Machair is not a truly natural habitat, in that it has evolved with man in all the machair areas, from the times of the earliest human settlement to the present day. Grazing is absolutely essential to prevent the machair becoming rank grassland and, while cultivation is not an essential criterion, it is worth noting that the best systems for nature conservation are those where traditional, rotational cultivation is still actively practised. Machair is restricted to the north-west of Scotland and the north-west of Ireland, and is found nowhere else in the world, though four biologists recently announced that they had found it in New Zealand[55]. Claims that it exists in the Falklands can be discounted on the grounds that rainfall there is too low. However, New Zealand is probably the part of the southern hemisphere which best matches the European machair areas in terms of climate and glacial history.

Machair systems probably occupy a total world area of less than 50,000 ha, so that the high proportion of it which has conservation designations is more than justified.

Stewart Angus, June 1977

Figure 4.20. Close-up of shell sand, Seilebost, South Harris

The Irish conservationists Bassett and Curtis[6] have pointed out that the distribution of machair in Ireland corresponds very closely with the Irish Gaeltacht (Gaelic-speaking area) and if one ignores the possibility of machair in the Northern Isles and the north coast of the Scottish mainland, then the same is probably true of Scotland. The human history and associated land use of the two main areas was very similar up till the time of the Crofting Act of 1886 (which applied only to Scotland) and it is entirely possible that a cultural definition of machair could be justified.

So how was this distinctive landscape formed?

As sea level rose with the melting of the last glaciers, the offshore moraines, eskers and terraces were swept away by the rising waters. Professor Ritchie has pointed out that sandy moraines occur down to sea level in the Uists and at depths of more than 70m offshore, but they are absent from a zone 21–29 km wide and up to 70m deep off the Uists, suggesting that the sediments have been removed from this area by wave action, possibly in association with rising sea level[41].

Sea level did not rise uniformly, but in stages, and also fluctuated. The shelled animals which grew on this shallow seabed contributed their shells to the existing mineral sediments, the proportion of shell rising with time, and the mixed sands were gradually transported shorewards, to be deposited on a series of beaches (depending on the prevailing sea level) and dried out by the wind.

There are three main components of beach sand: shell fragments, siliceous fragments such as quartz, and heavy minerals which are often dark in colour. Patches of pink sand on a beach may be garnet sand or perhaps tiny shells – sometimes predominantly of one species such as the Small Rissoa *Rissoa parva*, while darker patches may be grains of magnetite or hornblende. These different materials have different shapes and densities, so they will be transported differentially by wave action, leading to these recognisable patches but also affecting their subsequent deposition by wind. Dry, exposed sand will begin to move at wind speeds of 4.44m/s (16km/h)[36], so that all of the main machair coasts are potentially susceptible to windblow for more than half the time (using present-day information about wind – see Chapter 7). The sand grains move by 'saltation' (little jumps) which become longer as the wind speed increases. As the beach sand accumulated and dried, it was blown inland to build a line of coastal dunes and sand hills, probably by a date around 7,000–8,000 BP[47,19].

Ritchie and Whittington[49] compared radiocarbon dates for four apparently similar machair sites: two on Pabbay in the Sound of Harris, one at Cladach Mor between Tigharry and Scolpaig, and the other at Borve, Benbecula. The earliest discernible sandblow was around 7810 BP near the jetty on Pabbay, but the four sites had "substantially different" dates for the onset of sandblow. Comparatively few sites have so far been investigated in this manner, and the dates might have been related to different phases of machair formation determined by local topography.

This early sand was blown inland from beaches, dunes and sand hills to blanket the rocks and till-based topography left after the passing of the glaciers. Indeed, studies have revealed that the landscape underneath the machair is identical to that of the flat loch-studded peatlands which lie to the east[47]. On the flat, low-lying surfaces of the Uists and South Harris, the sand was blown far inland, forming an extensive sandy platform of *high machair*, with an eroded area between the high machair and the dunes which broadened with time, eating into the high machair surface, with sand being blown from there on to a succession of layers on the high machair as it crept inland, as shown on the model developed by Professor Ritchie (Fig. 4.21). In some places, the sand accumulated into tall sand hills: those at Luskentyre Banks are between 30 and 35m

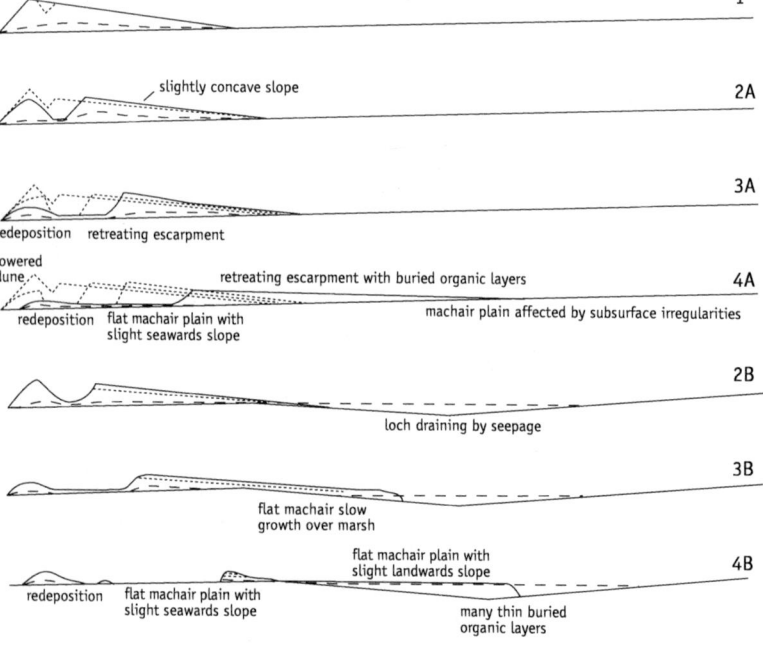

Figure 4.21. Two models of machair plain development (after Ritchie 1979).

high, and the highest in the Outer Hebrides[48]. Though we do not know anything about the level of the water table when the high machair was formed, the high machair surface is certainly well above the water table today and, being drier than the low machair, is more vulnerable to erosion by the wind.

While the role played by wind in the transport of sand grains has already been noted, it is most important to consider the importance of water, both on the surface and below the ground as the water table, as only dry sand moves readily. In our modern climate, at least, the driest times tend to be in early summer, before the sand has been stabilised by seasonal growth of vegetation: sand would blow inland, and stop when it got wet – on encountering lochs or marshes. Also, erosion would stop once sand had been removed to the level of the water table, where the sand is kept damp. Seasonal differences are such that areas of machair which appear very dry in summer may be flooded in winter; the water table also fluctuates seasonally (Fig. 6.1). The landscape produced by this

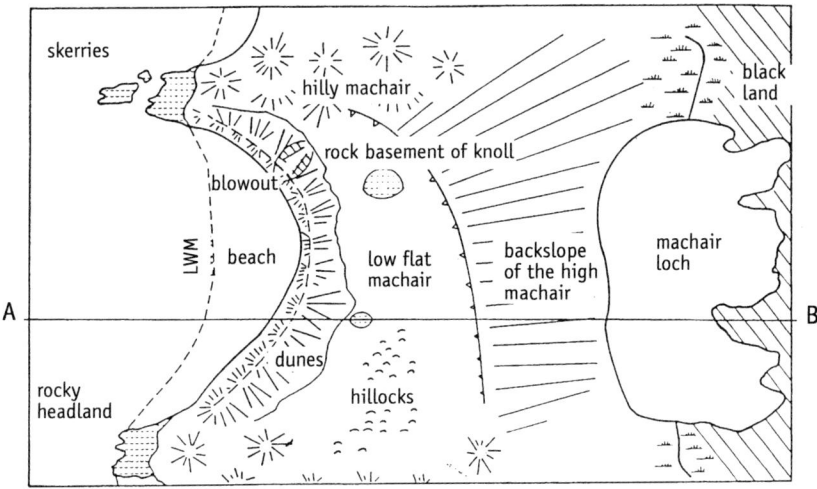

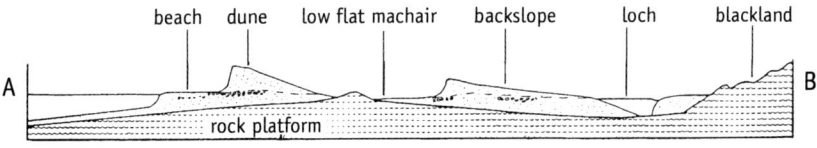

Figure 4.22. Diagram and profile of a typical Uist machair (after Ritchie 1971).

Figure 4.23. Machair mesa, Tràigh Eais, Eoligarry

type of machair formation was thus: a ridge of dunes fronting the sea, followed inland by a hollow, enlarged with time, down to the water table, at the back of which was the eroded cliff of the retreating high machair surface, which sloped gradually downwards inland as the supply of sand diminished with distance from the sea (Fig. 4.22).

The second type of machair formation identified by Ritchie in his model is similar, but instead of being deposited as a plain sloping gradually inland, the blown sand was deposited on the seaward side of a loch, gradually filling in the loch from the west, and forming a new, flat machair at the level of the loch's surface. This type of deposition continues to this day, albeit at a slower rate, in machair lochs such as Loch Bee, Loch Stilligarry, Loch Bornish, and in marshes on Newton machair and Daliburgh machair[47].

The modern machair plain is, on average, seven times as wide as the dunes (where they still exist), so that the ratio of grassland to dune is much higher than in 'links' on the mainland[47].

Ritchie's models of machair formation are essentially synoptic, in that most machair systems are hybrids between the two types, with complications arising from local (perhaps man-induced) erosion and secondary deposition. The high machair has disappeared from many areas, or is reduced to isolated remnants or 'mesas' which stand a metre or more above the present machair plain, often eroded on all sides with slumping of the edges – as distinct from secondary (and more recent) deposition, which tends to take the form of more rounded, completely vegetated hillocks and mounds. A remarkable mesa exists at Eoligarry, though little remains of its summit (Fig. 4.23).

Recent studies by the University of Sheffield[19] have confirmed the Ritchie models, in the identification of *stratified machair*, with discernible layers of soil and sand indicating a succession of stable periods (when soils formed) and erosion (when soils disappeared or were covered by blown sand from elsewhere), exactly as described by Ritchie (Fig 4.24). The darker layers representing soils can often be seen in eroded faces (Fig. 8.3), with the darker material revealed by microscopic studies as including organic material (e.g. humus) and even discernible animal dung. Radiocarbon dating and correlation with archaeological features of known date suggest that the stratified machair is a comparatively recent phenomenon, being particularly young on the Eoligarry isthmus in Barra, where the stratified machair is thought to date from the early 20th Century[19].

The Ritchie model depends, however, on an appreciation of the critical importance of sand supply, in that the phase of significant input of sediments from the sea is over, and sand movement today is mainly within the system – in a word – reworking. New input from the sea would result in ridges of dunes advancing seawards, for which there is little evidence[47]. Ian Mate[35] has questioned this conclusion, maintaining that the sand supply continues to grow with the provision of additional shell material from the seabed. In a later publication

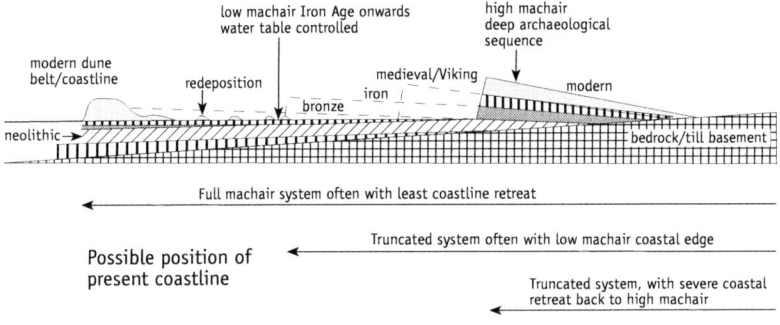

Figure 4.24. Sequence of machair development through time, related to specific archaeological periods (after Ritchie 1979).

with Graeme Whittington[49], Ritchie did not rule out Mate's argument, but contended that more work would need to be done on the evaluation of the different sediment inputs to machair sand at different times before a conclusion could be drawn.

A tentative chronology for machair evolution based on stratigraphy from archaeological digs at Northton[16], Udal[11] and Rosinish has been devised by Ritchie[47]:

1) The onset of some form of sand deposition as early as 3,750 BC.

2) Major primary deposition occurred in the period before the Neolithic II, i.e. around 2,500 BC.

3) Major primary deposition continued after periods of stability, or, a major phase of sand migration landwards followed until the phase of stabilisation in the 'Beaker Periods' around 1,750 to 1,500 BC. There would be minor episodes of disturbance and redeposition during this period.

4) Between the end of the Beaker Age and the beginning of the Iron Age there was a phase of redeposition on the higher, inland areas with material being derived from the deflation processes which produced much of the present day lower flat machair plains.

5) Before, during and just after the Iron Age there was a considerable number of relatively short periods of stability alternating with erosion and redeposition.

It is possible that this sequence was linked to a period of sea level rise lasting until 4,000 or 3,000 BC, followed by a period of slower sea level rise accompanied by the sea encroaching on the land. As will be shown in the next chapter, there is more than ample evidence for the latter, in the form of offshore beds of peat and even tree stumps submerged beneath the sea (see also Chapter 3). It is this period, after about 3,000 BC, which saw the formation of the drowned landscapes so characteristic of the coastline of the Western Isles.

Evans[16] used snail shells to suggest that there was a phase of woodland between Neolithic I and Neolithic II, but this has subsequently been shown to have been unjustified[1] and it is unlikely that any part of the Hebridean machair was ever wooded.

The machair grasslands are extremely low-lying, and all but the highest parts tend to flood in winter, or at least become very marshy. The altitude of any surface in a machair system seems to determine the habitat. Slopes can be very gentle almost to the point of being imperceptible on the ground, and even slight variations in elevation or local topography can lead to fundamental changes in

Stewart Angus, June 1992

*Figure 4.25. Machair plain sloping downwards from the sea (right) towards a
saltmarsh (left) on Vallay, North Uist. Further east, beyond the left of the
picture, the saltmarsh gives way to sand flats.*

habitat. Generally speaking, the higher areas are dry machair, slightly lower is
the low machair which in slight depressions will flood in winter. Slightly lower
comes marsh, then fresh water loch (Fig. 4.25). If the loch is connected to the
sea, it may be brackish, and the level of the connection with the sea will govern
the salinity by controlling the amount of sea water inflow. With a more open
connection to the sea, low machair vegetation can be displaced by saltmarsh
vegetation, and an even lower level will result in sandflats. All of these habitats
can occur within a single machair system, and there can even be variations within
each habitat. This variety of habitat in such a small area, ultimately controlled
by tiny differences in altitude, makes a major contribution to the nature
conservation interests of machair systems, and this delicate balance is vulnerable
to changes in sea level.

The chronological sequence above is somewhat conjectural, but even the
most recent detailed work on machair strata has not found anything which
contradicts this sequence[20]. This recent work has revealed, however, that
attempts at correlations of apparently similar deposits, depths and sequences
between sites could be grossly misleading, and that the formation of machair is
probably even more complex than had originally been thought, though the

Ritchie model still provides the basic framework. Certainly there are archaeo-logical remains beneath machair which are thousands of years old, but I am not alone in being struck by the recent nature of some of the artefacts sometimes revealed by the erosion of what had appeared to be very mature machair. The capacity of machair systems to change with time either in terms of erosion or stability should not be underestimated, and it may be that we have been too ready to overlook the capacity of at least parts of the habitat to recover very quickly indeed if the circumstances are right ... but which circumstances are right? This question is considered in Chapter 6.

Some machairs have particular features which make them stand out from others. The Grogarry system in South Uist is one of the best for studying machair landforms, as there is not only a good sequence of the main structures, but also a range of secondary deposition features, where blown sand has accumulated on top of the machair surface[37]. Pabbay in the Sound of Harris has conical dunes (Fig. 4.26) which in terms of their size and number are unique in machair, certainly in Scotland; these possibly originate from the storm of 1697 (see Chapter 5).

Where hills lie close to the source of blown sand, the shell sand may be blown far up the slope, as on Pabbay and Berneray in the Sound of Harris and at Eoligarry, Barra; shell sand is found on slopes of up to 42° at Eoligarry[45].

Stewart Angus, August 1993

Figure 4.26. Aerial view of conical dunes on Pabbay, Sound of Harris

By 1994, twenty instances of intertidal peat were known from the Uists alone, most of them associated with machair[49]. These and the many archaeological sites are of vital importance in establishing a machair stratigraphy in the study of machair evolution.

The organic layers (usually peat but sometimes wood) which are so often associated with the beaches of machair systems and which are now below High Water Mark provide clear evidence of coastal retreat, but as neither peat nor wood is usually closely associated with machair, these deposits pose additional questions about machair formation. Ritchie and Whittington[49] have provided a model of the origin of intertidal peat, based on their work at Cladach Mor. Their theory depends on the existence of topographic basins relatively near the coast in which organic layers could accumulate (or, exceptionally, in which trees could grow) prior to the onset of sand blow, and when sea level was lower than today (Fig. 4.27). The circumstances they describe are entirely compatible with our knowledge of sub-machair topography and former sea levels.

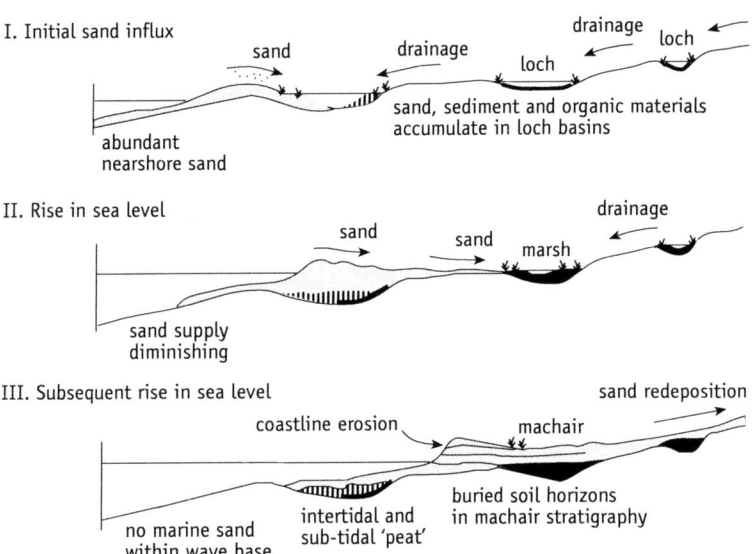

Figure 4.27. Development of organic layers in machair
(after Ritchie & Whittington 1988)

HUMAN INFLUENCE ON COASTAL DEVELOPMENT

This section deals with influences other than coastal protection; sea walls and other devices installed to combat erosion are discussed in Chapter 6.

Stewart Angus, March 1987

Figure 4.28. Aerial view of Stornoway airport spit

The northward-projecting spit just north of Stornoway Airport (Fig. 4.28) owes its origin not to marine longshore drift but to the movement of fine sand by the wind. The spit is a relatively recent phenomenon, and its development may have been triggered by land reclamation and the building of a boundary wall during the original construction of the airport[48]. It is possible that the recent extension of the main runway will also have a significant effect on this dynamic system.

Until the Second World War, the area between Melbost and the sand flats, identified on the 1898 Ordnance Survey map as the 'Carse of Melbost' (Fig. 4.29), was common grazings, also used by the Stornoway Golf Club for their course (Fig. 4.30). In 1933 the Club agreed to permit aircraft to land on their course, which inevitably led to the construction of the airport on the course in 1939. Though the Club attempted to redesign their course to fit in with operations until May 1940, they took advantage of compensation from the Air Ministry to move to the present site in the Castle Grounds in 1947[32]. The old course was a magnificent links, but it had hazards from cattle, mobbing terns, and even the incoming tide, which would flood the first fairway during high spring tides[21].

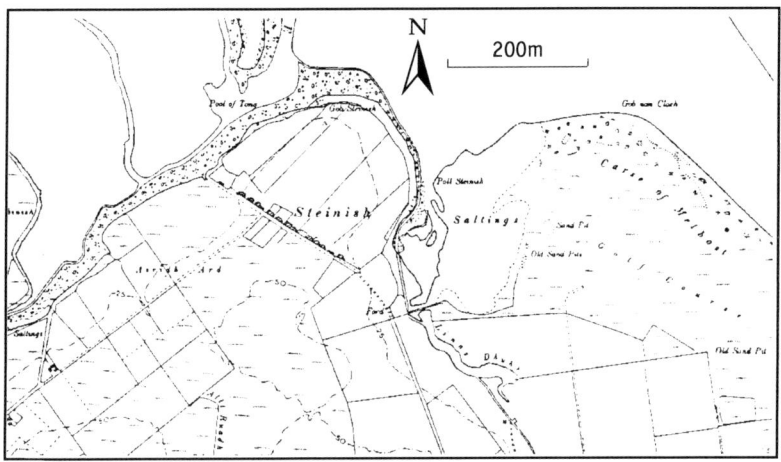

Figure 4.29. 1898 Ordnance Survey map of the area now occupied by the sand spit north of Stornoway Airport. Note the complete absence of any spit. Reproduced by permission of the Trustees of The National Library of Scotland.

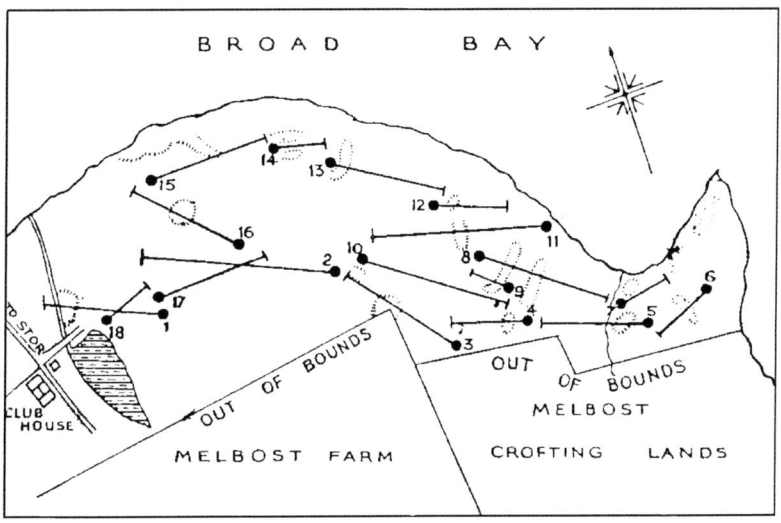

Figure 4.30. Layout of the old golf course at Melbost, now occupied by Stornoway Airport. The promontory on the right is a rocky headland, beyond the margin of the sheet from which Fig. 4.29 was taken. Reproduced by kind permission of Stornoway Golf Club

There is no sign of any peninsula on the old maps – the coast followed a fairly smooth curve from Steinish round to Melbost Point, and the spit is believed to have grown on rubble dumped there during the War. This is the only true sand spit in Lewis: all the others are dune systems based on a shingle peninsula. I have tried and failed to obtain aerial pictures from the RAF showing the wartime airfield, but the Sandwick Historical Society showed me one they had acquired – from the Luftwaffe, taken in 1942 (Fig. 4.31). A fair amount of rubble seems to have been used to stabilise the coast during or following construction, leading to a slight change in the coastal outline. As is so often the case, this rubble is now becoming exposed by erosion. Ritchie and Mather[48] believed that most of the sediment which built the spit came from offshore, and modern aerial views display a fine series of hooks, together with a few circular dunes near the tip (Fig. 4.28). Though shell fragments constitute only 16% of the sand material here (Table 4.1), I have seen astonishing amounts of dead shells of the Banded Wedge Shell *Donax vittatus* on this beach, in extensive piles up to half a metre deep. The additional shelter provided by the growing spit encouraged the extension of the saltmarsh, and the 'new' area of saltmarsh is

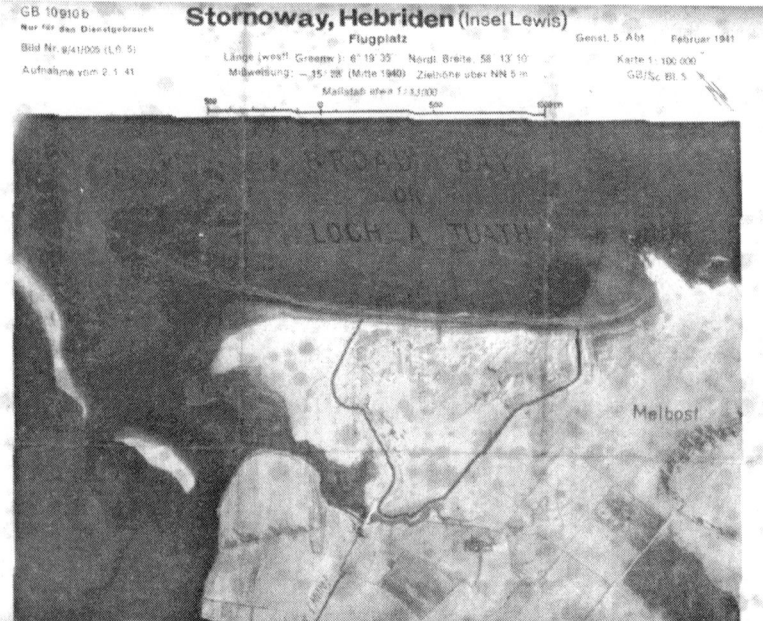

Figure 4.31. Aerial picture taken by the Luftwaffe during construction of Stornoway Airport in 1942. Reproduced by kind permission of Reg Brown.

probably the sandiest area of saltmarsh in the Outer Hebrides, in marked contrast to the older, muddier systems nearby, and the new system is probably still growing. This area also came in for attention for land reclamation following the Second World War, when the need to import milk from the mainland was regarded as a major problem. In its first Report, the Lewis Association advocated the enclosure of the 'Cockle Ebb' by building 'sea dykes' of about 350m, in two sections, which would then enable an area of 174ha of new land to be created, with a further 120ha improved in the process[38]. The *Stornoway Gazette* of 3rd March 1959 reported that the reclamation scheme had been abandoned following an unfavourable decision in the Land Court.

Beach	Grid Reference	Location	%CaCO$_3$
Barvas	NB343508	HWM (over shingle)	41.70
Barvas	NB344508	young machair	48.30
Barvas	NB359517	old machair	44.30
Branahuie	NB475325	HWM	7.50
Coll	NB464385	HWM	46.70
Dalbeg	NB226458	HWM	46.70
Dalmore	NB215451	HWM	36.20
Eoropie	NB512645	HWM	56.40
Garry	NB536498	HWM	52.80
Gress	NB493413	HWM	61.90
Mangersta	NB008308	HWM	38.40
Melbost	NB460340	end of spit	16.00
Swainbost	NB505637	HWM	45.60
Tolsta	NB543485	HWM	50.30
Tolsta	NB537493	stream mouth	46.70
Tràigh na Berie	NB106358	HWM	47.00
Tràigh na Clibhe	NB083365	HWM	23.20
Tràigh na Clibhe	NB083364	stream mouth	20.90
Tràigh Valtos	NB097367	HWM	54.50
Camas Uig	NB049329	HWM	39.40
Camas Uig	NB049328	machair	36.20
Aird Nisabost	NG050970	quarry	50.60
Borve, Harris	NG030950	blowout	50.90
Corran	NG064980	machair	45.70
Hushinish	NA990120	blowout	82.90
Luskentyre	NG065995	dune	44.40
Luskentyre	NG065995	HWM	53.70
Scarasta	NG005935	beach	67.80

Table 4.1. Calcium carbonate content of shell sand, Lewis and Harris (above, from Ritchie & Mather 1970), and the Southern Isles (overleaf, from Ritchie 1971)

Township	Island	Site	%CaCO$_3$
Ahmore	North Uist	island	52
Baleshare	North Uist	machair	32
Baleloch	North Uist	machair	68
Balelone	North Uist	machair	47
Berneray	Berneray	machair	54
Borve	Berneray	sandhill	53
Balranald	North Uist	machair	48
Balranald	North Uist	beach	63
Balmartin	North Uist	machair	35
Carinish	North Uist	strand	23
Clachan Sands	North Uist	machair ridge	38
Grimsay	North Uist	strand	1
Hosta	North Uist	sandhill	77
Hosta	North Uist	machair	80
Hosta	North Uist	beach	74
Kirkibost	North Uist	strand	35
Machair Leathann	North Uist	machair	74
Paiblesgarry	North Uist	strand	36
Sollas	North Uist	machair	67
Trumisgarry	North Uist	strand	44
Dunganachy	Benbecula	machair	37
Aird	Benbecula	beach	58
Balivanich	Benbecula	airfield	30
Nunton	Benbecula	beach	55
Gramsdale	Benbecula	strand	22
Uachdar	Benbecula	strand	34
Lionacleit	Benbecula	beach	28
Lionacleit	Benbecula	dunes	19
Borve	Benbecula	machair	6
Askernish	South Uist	machair	64
Bornish	South Uist	machair	52
Bornish	South Uist	beach	48
Daliburgh	South Uist	machair	30
Gerinish	South Uist	machair	62
Garynamonie	South Uist	machair	62
Drimsdale	South Uist	dunes	60
Drimsdale	South Uist	machair	53
Howmore	South Uist	machair	55
Eochar	South Uist	machair	84
Eochar	South Uist	low machair	53
Eochar	South Uist	high machair	40
West Kilbride	South Uist	machair	39
Milton	South Uist	beach	74
Stoneybridge	South Uist	beach	42
Stilligarry	South Uist	machair	54
Smerclate	South Uist	machair	60
North Boisdale	South Uist	machair	49
Allasdale	Barra	machair	83
Allasdale	Barra	high machair	62
Eoligarry (jetty)	Barra	sandhill	50
Vatersay	Vatersay	machair	78

Another system which has changed considerably is Scarasta. Pictures taken in the early 1950s seem to show a bare, sandy beach where today there is a group of established dunes, several metres in height. These are thought to have resulted from the accumulation of sand on driftwood just after the Second World War, when the build-up of sand was so rapid that a fence was buried within five years of erection[9]. Once marram became established, the growth trapped more sand, and the dunes grew, as they are probably still growing.

From comparison of the series of maps of the Uists, it is clear that major dune stabilisation exercises and loch drainage were carried out. The sources cited in the next Chapter describe some of this work, but much of it is undocumented. Alexander (or Alasdair) Macleod, who from 1809 was Factor of North Uist, and later became involved in land management in South Uist and Benbecula, is often credited with the greater part of this work. He also worked as a physician, being known as 'An Dotair Ban' (the Fair Doctor), and seems to have been held in very high regard by local people as both a doctor and a land manager. The Reid Map of 1799 of North Uist[39], however, and the Bald maps of 1805 of Benbecula[4] and South Uist[3,5], confirm that most of the major drainage work had been completed before Macleod arrived in the Uists.

Beveridge[7] reported that the proprietor drained the fresh water Loch Paible in North Uist in 1793. The loch had been separated from the sea by a sandbank. This was breached, with the intention of draining the loch and reclaiming the land for agriculture. Instead, the sea entered the loch, and created the inlet seen today (Fig. 4.32).

Stewart Angus

Figure 4.32. Loch Paible, North Uist

An Dotair Ban did, however, carry out at least three major improvements which he described in a paper in the *Transactions of the Highland Agricultural Society* in 1831.

Around 1824, Macleod decided to drain Loch Mor on the island of Boreray in the Sound of Harris, despite finding that the bottom of the loch was only two and a half feet above Low Water Mark. Using explosives, he excavated a drainage channel to the sea almost 300m long through the shingle barrier, providing a flood gate to prevent the entry of the sea at high tide. This work drained the loch within two days. He then covered the part of the drain that was on the shore and erected an embankment 1.6m high to protect the loch from the highest waves. He thus reclaimed about 24ha of good quality land, and the cost of the works was repaid by the value of less than two years' crop[33]. Despite the early success of this scheme, the reclaimed land was regularly flooded by the time that the Royal Commission was taking evidence in 1883, due to lack of maintenance of the drain and flood gate by the estate[51].

Stewart Angus, August 1983

Figure 4.33. Loch Mor, Boreray

Macleod also described the drainage of a loch which was located between Kilphedder and Scolpaig in North Uist, also by creating a channel to the sea. There are three lochs in this area, but the Reid map of 1799[39] seems to show exactly the same three lochs which are there today. The 1799 map shows the two northern lochs linked as one, and it is possible that Macleod drained this hourglass-shaped loch but that neglect of the drain led to the restoration of most of the former level within a few decades, as on Boreray.

His third scheme was even more ambitious, and involved the building of sand embankments at Gearradubh, on the north coast of Grimsay NF8557, in order to reclaim land from the sea. Alas, there is no sign of his work today.

In 1967, the Highlands and Islands Development Board (HIDB) commenced experimental bulb-planting in North Uist, with 2.4ha of bulbs planted in consultation with experts from the Netherlands. By 1969, this had been increased to 13ha of crocuses, tulips and daffodils at Balemore and Kyles Paible, and was believed to be technically viable, with yields comparing favourably with those obtained in the Netherlands, but it was thought that in order to make the project commercially viable a much larger area would be required. In 1968, they commissioned engineers from the Netherlands to investigate the feasibility of reclaiming 600ha of Vallay Strand. Having concluded that this would be both possible and economically feasible, they requested the authority of the Secretary of State to proceed with the reclamation. By 1971, however, reports were not quite so favourable; the application to reclaim the intertidal area was withdrawn, and the scheme was abandoned completely in 1972. During this period, the HIDB also examined the feasibility of land reclamation of intertidal sand at Baleshare, but did not proceed with this[24].

Machair is important for a range of interests: landscape, archaeology, landform evolution, plants, birds, invertebrates, and the history of human land use and settlement. None of these can be looked at in isolation: they interrelate, and the bird interest in particular is totally dependent on crofting land management. Much work remains to be done in all fields, but biologists, archaeologists, geographers and historians are now working in close co-operation to try and throw more light on this fascinating and (in world terms) very restricted habitat.

The machair was created by the wind, and the wind is one of a range of factors now destroying this vulnerable coastline. Because of the great interest in this subject within the Western Isles, I have devoted two Chapters to machair erosion, the first of which (Chapter 5) looks at the historical record.

5

HISTORIC EROSION

It was the kelp industry which had doubled the rents. When the kelp industry ceased, due to no fault of the crofters and cottars, they felt that in equity they were entitled to pay only half the rent. The proprietors, after losing their profits from kelp, had now to depend on the payment of the rents and were therefore not anxious to diminish it. They were all the more anxious that it should be regularly paid and when arrears appeared they lent a keener ear to the blandishments and offers of ambitious improving farmers. The latter had no use for crofters and wanted them cleared. A series of very bad harvests really from 1836 to 1851 combined with the failure of the potato crop in 1845 and the following years was a disaster for the crofters. (Morrison 1982)[27].

I have departed from custom in this section by quoting large passages verbatim (retaining original spelling) from the more useful historic sources to set the scene for the next chapter, which examines erosion as a contemporary problem. My comments within the quoted sections are given in square brackets. The quoted sections are distinguished from my own text by indented margins.

The various islands and groups of islands are dealt with in a south-to-north order, though many passages have a wider geographical application.

The main sources cited are: the *Old Statistical Account* (OSA)[30] of the 1790s, Macdonald's *General View of the Hebrides or Western Isles of Scotland* of 1811, the second or *'New' Statistical Account* (NSA)[29] of the 1840s, and the work of Erskine Beveridge[5] on North Uist. Within each island, the sources are quoted roughly in the chronological order of the events to which they refer.

The passage quoted at the beginning of the chapter might appear unrelated to erosion. In fact, in the closing years of the 18th century, until around 1822, the years of the kelp boom, interrelated economic activity, population growth and agricultural intensification put immense stress on the machairs, but this was insignificant compared with the stress of the later consequences for the inhabitants, as the history of kelp, which has many links with the machair coasts, played a pivotal role in the events which led to the Clearances, as described later in this chapter.

It is interesting to note that neither of the first two main works on the Western Isles – by Dean Monro and Martin Martin[23] – refer to machair erosion in any form. Given that they were closely connected with landowners, and landowners were extremely concerned with land as it was the source of their income through rentals, this might almost be taken to mean that erosion was not regarded as a problem at the time of their visits. Martin is believed to have visited the Uists in 1695, just before the great storm of 1697, but makes no reference to the loss of Hussaboste – was it too long ago to be of relevance? Though I have quoted *Macfarlane's Geographical Collections*[20] this work, believed to have been written in the first half of the 17th century, contains some bizarre information which can best be classed as mythology; the paragraphs cited below seem to be compatible with the historic record, but they should nevertheless be treated with caution.

Archaeological sites such as Rosinish in Benbecula[35], Udal in North Uist[11] and Northton in Harris[13] reveal that there were certainly phases of stability and sandblow well before historic records began. Indeed, they show an onset of sand deposition as long ago as 5,700 years ago, with even older dates for Pabbay in the Sound of Harris (see Chapter 4). Other archaeological sites on the machair plain are underlain by blown sand, and some are now waterlogged or on an eroding shore[33].

All of the 'acres' cited in this chapter are likely to be 'Scotch acres' rather than Imperial acres. Each Scotch acre is roughly equivalent to 1.27 imperial acres, or 0.514 hectares, so that an approximation of any area in hectares can be obtained by dividing the Scotch acre figure by two. (A hectare is 2.471 Imperial acres). Any areas given in figures in this chapter should, however, be viewed with extreme caution.

GENERAL

It must not be imagined that this Hebridian sand is barren soil, it being destitute of vegetation only when drifting loose. When in some degree fixed by moisture, or the interspersion of pebbles and shells, it affords excellent crops of barley, when manured with sea-weed, and its natural pastures are by far the best of the Hebrides. (Macgillivray 1828)[21].

BARRA

The Main seas doth come from the West, and the other seas from the east, and almost the saids two seas doth forgadder and meet with other. And they have cutt and broke the lands in divyding the Illand of Barray into two parts almost next to the litle Chappell of Kilmore. (Macfarlane's Geographical Collections c.1630)[20].

The depredations of the sea and storms upon [Barra, Uists and Benbecula] are alarming, and indeed incredible to those who have seen nothing similar. It is probable that the country in many places, has lost one-fourth of a mile of its breadth, or 600 Scotch acres by the sand drift and encroachments of the sea during the last two centuries. (Macdonald 1811)[19].

On this west shore however, there are several Tracts of very good light Soil, which afford crops of Bear and Oats, but greatly prejudiced by the extensive Fields of blowing Sand, which turn and wheel, and move over the Country, in a very hurtfull way. This Sand Drift, against which they have no Defence, has oblidged them to remove Houses, and even entire Villages, in several parts of the Island. ... On the 19th February 1749, a Hurricane from that Quarter [SW], with a high Tide, broke over for the first Time, an Isthmus which divides the Island in two parts. The Isthmus was very extensive, and consisted of excellent Land, but ever since that Inundation has been a blowing Sand, though the Sea has never again forced its way over it. ... It is but 12 years, since Potatoes were first cultivate in Barra, and without them, they would not now be able to support themselves in Grain, since a great deal of their best Soil, has of late years been rendered unserviceable by the Progress of the Sand Drift. (Walker 1764,86-88)[36].

[Lady's Bedstraw *Galium verum*] grows most luxuriantly and affords the largest Roots upon the sandy Downs, so that it is found in Barra in the greatest Perfection; but in these Places, the People are prohibited from digging it up [for dye], for as there is nothing but blowing Sand under the Grass, when once the Surface is broke a Sand Drift commence, which will soon destroy a whole field. (Walker 1764,90)[36].

The soil in general is thin and rocky (excepting the northern end, which is a mixed soil, and almost free of rock); there is also a great deal of sand, which is blown one way or other with every gale of wind, so that a great part of the best cornland has been thus blown away, or covered with sand. There is some meadow ground between the hills. The ground here requires that the manure be thick laid, in order to produce a tolerable crop ... Seaweed is the principal manure here; as that is sometimes precarious, the crop must be so also, for when a sufficient quantity of seaweed is not cast upon the shore, a plentiful harvest is not to be expected. Formerly the sea-weed that grows upon the shore was used for manure; but since kelp has become so valuable, the proprietors every where have restricted the people from cutting it for that purpose, which is currently prejudicial to agriculture (OSA,138)[30].

... the sand is driven to and fro by every successive gale, leaving nothing behind but the rocky skeletons of granite or whinstone, where once the ground was clothed with a beautiful variety of red and white natural clover (NSA)[29].

Another great obstacle to the improvement of their lands is the manufacture of kelp, at which the people are bound by their holdings to labour during the summer season, and even sometimes to the end of August, the fittest time for the improvement of their lands, and attending to the management of the crop in the ground (NSA)[29].

Recent work on the history of the Eoligarry isthmus[15] has revealed a fascinating range of materials buried in the sand which can be used as stratigraphic markers, including rabbit traps, plastic bottles, and coal from a shipwreck, all of which can be dated within a narrow time span. This work has established that the present coastal dunes date from the end of the last century, and that prior to this, there were two projecting areas of land, joined by a low-lying area which was occasionally flooded during storms.

PABBAY, BARRA

Pabay ... is greatly spoiled by the Sand Drift. (Walker 1764,85)[36].

One of the islanders ... also mentioned that till a few years ago the plateau was covered with grass, and when the peat on the island was used up they cut the turf for fuel, and the wind blew off the underlying sand, which was about six feet deep, and exposed the old surfaces containing the middens. (Wedderspoon 1912)[37].

SOUTH UIST

... the Atlantic has formed a sort of barrier to itself, by heaping up sand and stones in many places, so as to resemble the work of human industry. The barrier is indeed insufficient, and is gradually retreating from its mighty opponent, but at high spring tides, with westerly storms, it often saves the most valuable part of the country from being inundated and overwhelmed. This holds in a remarkable degree of the southern end of the island of South Uist, from the march of Boisdale to that of Ardmichael in the middle of the district. (Macdonald 1811, 786)[19].

The remaining 51,720 Acres [of South Uist] may be deemed irreclaimable, being composed of steep rocky Mountains, and extensive Tracts of moving Sand. ... The West Coast of South Wist for the Extent of 33 Miles is a dead Plain; in most places, about 2 Miles broad, and of so deep a Sand, that there are no Springs in it. ... The Shore is fenced with vast banks of blowing Sand, with which the whole Country is flooded in Time of Storms. This Sandy Deluge is of the utmost

Detriment here, as it is indeed along all the West coasts of the long Island, and no effectual means have ever been used to restrain its Devastation. In this Island, it has come with such Violence upon many Houses and Villages, that they have been obliged to be removed, to escape being overwhelmed. It has choaked up Rivulets, and made the Waters of the Lakes from whence they issued, advance upon the Land, and overflow some of the best fields in the Island. The Crops also, are in the greatest Danger from it, and it was indeed Melancholy in the beginning of last August, after a Tract of high Winds, to see some excellent Fields of Bear, turned in a few days into fields of Sand, in some places a yard deep. ... In South Wist the Foundations of Stone Walls are to be seen at the lowest Ebb, above half a mile from the present Floodmark. (Walker 1764,74-75)[36].

Towards the west side of the parish, the soil is totally light, and perfectly sandy, and the most part of it rendered quite useless, by the severity of the constant storms, that blow from the W. with the force of the sea, during the winter and spring seasons ... As the soil to the west side of the parish is for the most part light and sandy, it of course must be barren of itself, without the force of manure. There are delightful fields to be seen covered with the finest natural grass in summer; yet, in the winter season, many of these very spots are covered over with drifted sand in such a manner, that the least trace of verdure cannot be seen for many months (OSA 127)[30].

[Howmore] farm is mostly under sub-tenants who are very industrious, they came there from North Uist and showed the fast example of reclaiming the blowing sand and bringing it into pasture. ... [Ormiclate] farm has been considerably improved by the present tenant, particularly in bringing the drift sand along the shore into cultivation, he has already cropped about 40 acres and has got the surface fixed so that it will soon be good pasture. ... [Daliburgh and Kilpheder] towns are under small crofters and the ground appears to be exhausted. The machar has been very improperly broken up and cropped, and the banks are now in a dangerous state and exposed to the winter gales. (Fleming 1839)[14].

However, though late, this sand-drift was in great measure remedied in several places, within the last twenty years, as far as was practicable. In five or six farms, the sand-banks which were repairable, were, at great expense, covered with green sod, taken from the neighbouring ground, and in the course of a year, defied the strongest gales, and which, but for this prevention, would soon have been entirely blown away, and have left a barren desert; whereas, they are now covered with fine grass bent (*Arundo arenaria*) [marram] and afford both shelter and food to cattle. Another method has also been tried, on a small scale, to prevent sand-drift. Bent, or sand-grass, has been planted in two farms in the Dutch style, and found to answer the purpose exceedingly well. This bent is used by the parishioners in making sacks for carrying their corn to the mill, and for horse graith, etc. There is a large sandy tract near the northern extremity of South Uist, named *Machare mianach* [west of Loch Bee] consisting of about 600 acres, which, for the last hundred years, has been of no use whatever, now partly brought into cultivation, being contiguous to abundance of seaware occasionally cast ashore. There were about 150 acres of it under bear or bigg last year ... If this ground, tilled last spring, were allowed to remain without crop for a few years, and if the adjoining part of 450 acres were brought into cultivation in succession in the same way, the whole (now a complete waste) would, in the course of a few years, be a delightful piece of ground, yielding abundance of fine grass. (NSA, 191-192)[29].

BENBECULA

There will be 8,000 Acres of arable land in [Benbecula], and not above 2,000 irreclaimable, which consist mainly of fields of blowing Sand. (Walker 1764,69)[36].

Hemmed in by croft fences, derelict wartime shelters and nearby houses, this ancient Benbecula landmark [Borve Castle] gives little away as to the role it once played over long centuries of island history. Unusual, if not unique, amongst other fortress dwellings of the same period, it stands not in a position of any apparent physical prominence but appears stranded in the middle of the low-lying undulating machair which stretches the length of [Benbecula's] west coast. Appearances are, however, deceptive. Borve did indeed once occupy a distinctive site, the traditional rocky islet perch to be exact. But the configuration of the land has changed, the coastline altered, even in the last 150 years as the details on older estate maps clearly reveal. A memory of this former topography is retained in local tradition. A clue lies in the name for the area of land which curves inland round the back of the castle. It is known as *an traigh*, the shore. There is even a tradition of seals being shot in the area.

In former days, Culla Bay was known as *Culladh Mhoire*, the treasury of Mary, so great was the quantity of tangles cast up on its shore, so valuable in the careful balance of working yet nourishing the land. ... Tenants were warned that the use of seaweed for manure could lead to prosecution. (Burnett 1986)[8].

Bald's map of 1805[4] shows "Toam" [An Tom – now airport] in Benbecula as "Blown Sand now mostly arable" and the Borve-Lionacleit area is shown as "blown sand". Ray Burnett's account[8] of the coastline around Borve demonstrates that not all the change has been loss.

NORTH UIST

There was ane Ancient man in a toune in Wist called Killpettill and this old man said that he was sex or seven scoir of years old and he did sie another Church with the lands of the Parish wherein that church did stand. And these lands were more profitable fertill and pleasant then these that are in Wist now. And that his father and mother, his grandfather and Grandmother did see another parish Church which was destroyed with the sea long agoe. And that they did call that Church Kilmarchirmore. The next was called Kilpettill, and this Church wherein he doth dwell now into, was called Killmony which is now called Kilpettill that is to say the Mure Church, because it lyeth next the Mures. Mosses and Mountains And this Church is below the sands except foure or fyve foot length of the pinnacle of that church And the airt of there houses which are nearest the seasyde for the Wind doth blow up the sand upon the lands and the churches were destroyed with the sea which were principall Churches of Ancient. Certaine of them will be seen when the sea ebbs in the summer tyme. And the Countrey people will take Lobsters out of the windowes of the Pinnacle of that which was first called Kilpettill before it was destroyed with the sea. (Macfarlane's Geographical Collections c.1630)[20].

The oldest men report this Isle [Uist] to be much empayred and destroyed be the sands ovirblowing and burieing habitable lands, and the sea hath followed and made the loss irreparable, there are destroyed the tounes and paroch churches of Kilmarchirmoir and Kilpetil. And the church of Kilmonie is now called Kilpetil, that is the church of the muir for so it lay of old nearest the muirs, but now the sea and the sands have approached it, there be sum remaynes of the destroyed churches yit to be seen, at low tydes or Ebbing water. (Macfarlane – c.1630)[20].

Mould[28] points out that "Killmony" is Gaelic, either the church of Mary (the Virgin) or of St Maelrubha and has nothing to do with the Scots word "Muir". Kilmarchirmore is perhaps the church of the great machair.

The gluten which [seaweed] contains tends also to consolidate and bind sandy soils, and to give them, in the course of a few years, if properly cropped, and not exhausted, an increase of valuable staple. This may be easily perceived as its effects in Tyree [Tiree] and in North Uist. The farm of Pebbill [Paible], in the last mentioned island, possesses land worth L.2. an acre, which has been converted from barren sand to its present state, approaching to the finest loam, by the constant application of seaweeds for a century past. The same effects were observed in Lewis and Harris. (Macdonald 1811, 407)[19].

This farm [Vallaquie] is much injured by sand Drift from the broken banks stretching towards Oransay, which are of great extent.

This farm [Sollas] has suffered dreadfully from the effects of sand drift, the whole extensive range of fine *Machar* pasture stretching from the Houses to near the point of Ard Vorran, being one sheet of white shelly sand, which has been nearly blown down to the sea level, and a great part of which must consequently be washed by the tide, in time of high floods.

There is also a considerable extent of valuable land (crofts 16. 17. 18. 19. and 20) of the use of which the tenants have been of late years almost wholly deprived by an accumulation of water, in consequence of the drain from it being filled up with blown sand, owing to the cause that I have already described. The land on which this water has lodged is naturally of excellent quality, and therefore the drain should again be opened to the sea and constantly attended to afterwards, as the loose nature of the sand renders such attention absolutely necessary. The expense of executing this drain properly should not exceed 2 or three years rent of the ground lost from the want of it.

There are some sand banks on the last two farms [Dunskellor and Middlequarter] that require attention also.

The quarter part of this Island [Kirkibost] is drifting shelly sand bearing no vegetation. (Maclean 1830)[22].

... in 1542 the valued rental of North Uist was officially reduced by two or three 'merk-lands' on account of encroachment by the sea, presumable at some time then quite recent, no particular locality, however, being specified.

Nearly two centuries later, history seems to have repeated itself, as we find a document addressed from North Uist in 1721 to the Forefeited Estates Commissioners, wherein "*We, the wadsetters, tacksmen, and possessors underscrivers attest and deliver – That in regarde of the extreme poverty reigning amongst the haill tennants and possessors within the Barony of North Uist occasioned by a murain in our cattle first in 1717 but more especially this year whereby a great many of our cattle have perished to the number of seven hundred and fourtie five cows, five hundred and twenty three horse, eight hundred and twenty sheep ... and moreover we attest and deliver that about Candlemas last the sea overflowed several parts of the countrie breaking down many houses to the hazard of some lives which have impaired the lands to such a degree as its possible it may happen more and more that they cannot answer to the worst sett in former times*". Again, no locality is given, but from the signatures to this 'attestation', the lands affected in 1721 evidently lay at the west and north shores of N.Uist, including the districts from Paible to Kilpheder on the west, and those of Griminish, Vallay, Dunskellor, and Boreray on the north. (Beveridge 1911)[5].

[Marram] is very tough, and in some degree elastic, and is used by the poor people for many purposes, such as mounting for their crooksaddles and creels, sacks for their corn, meal, etc. It makes excellent mats for doors and passages. But its principle use is, to plant it for the suppression of sand drift. A great and important improvement also has been for some years extensively carried on, in the suppression of the sand drift, and evil formerly of great magnitude in these parts. The suppression of the sand drift is effectually secured, by sloping the sand-banks, and covering them with sward from the neighbourhood; they are thus become firm, and produce grasses of the same kind as in the situations whence they were taken. Bent [marram] is also employed successfully for the same purpose. Very near the seashore, and on extensive sand-flats, the planting of bent is the best method hitherto discovered for the suppression of sand-drift. (NSA, 1837)[29].

On Reid's estate plan of 1799[31], considerable areas of machair are shown as "sand drift", and in a report compiled in 1799-1800 by Blackadder[7], the surveyor comments that the problem of sand drift "in some places may be prevented by sloping down the banks where they are broken and sowing artificial grasses"[9]. Ritchie[32] describes a high sand ridge which rises to more than 20m, now deeply dissected, in an area to the north of Loch Paible noted as as "sand drift" on the Reid map.

BALESHARE

The sand Drift has made great Devastation in many parts of North Wist, and continues yearly to be more and more formidable. Several parts of the Country which are but little raised above the ordinary Level of the Sea, have also suffered from extraordinary Tides, which are frequently occasioned by the great Violence of the South West Winds, combined at the full or Change with the heavy Swell of the Atlantick. Such a Tide, in the year 1756 broke over an extensive Isthmus, and turned it into a Heap of Sand which before would have pastured 100 Cows in Summer for a Fortnight or three Weeks. By this Irruption, the Peninsula of Inchemish [Eachkamish], which is two Miles long and a Mile and a half broad, was disunited from North Wist and turned into an Island, and by the breaking of the Isthmus a Deluge of Sand has been poured in upon the Farm Town of Ballyshar. The houses in this Village are now blown up to the Roofs, so that there will soon be a necessity of having it removed further into the Country. Near this place the Sand Drift has also choked up a Canal, which had been dug 7 or 8 Feet deep and half a Mile long in order to drain two Lakes which are now by that means destitute of Level. (Walker 1764, 64)[36].

The islands of Kirkibost and Ileray [Baleshare]... the former is 1 mile long, but very narrow; the soil sandy, lies quite exposed to the Western Ocean, which makes yearly encroachments, and is in danger of being soon blown away by the wind. (OSA 103-104)[30].

With regard to the earlier devastation of about the year 1540 ... we could conjecture that the principle damage was further south, – perhaps chiefly at Baleshare Island, where so recently as 1859 (July), through a high tide accompanied by a south-westerly gale, 'the soil was washed away and channels formed that had never existed before'. (Beveridge 1911, vii)[5].

The island of Baleshare (of which Ileray forms a part) must have been at one period (and that not very remote, a valuable sheet of fine Machar land, but like the adjacent Isle of Kirkibost, it now presents a very different aspect. A very large portion of it already having been destroyed by moving sands and the remainder constantly exposed to the same destruction.

It is hardly possible however to reclaim this extensive tract of desert by the ordinary means resorted to in the Long Island, because the devastation is so very complete that not a vestige of green sward, nor even a tuft of bent can be observed from one end of the dreary waste to the other – But I should hope that the future progress of the evil might in some measure be averted, partly by transplanting bent from other places to the inner border of the sand drift, or where it passes into the solid surface, partly also by the common process of covering with sod, but chiefly by prohibiting the Tenants from losening the soil, either by ploughing or any other mode of culture within reach of the Sand drift. (Maclean 1830)$_{22}$.

Evidence of John Macdonald, Crofter, Illeray, Baleshare Island [to the Napier Commission]:

... the best part of cropping and grazing ground has of late years been rendered next to useless by the encroachment of the tides, and in common with similarly situated places a large tract of it was completely carried away by the high tide of November 1822. We feel the want of this piece of ground very much, as it was, as stated, the best arable land we had.

The flood tide has injured us. Illeray is divided into sixteen crofts; eight at the one end and eight at the other. Those at the far end are quite destroyed by the sea. It opens on the ocean, and the Atlantic has broken in upon us. It takes away our manure and our soil and the cut corn and the potatoes. It drowns the sheep and spoils the fresh water, so that our cattle are being injured by it, and we cannot cultivate the ground because it is continually being flooded. It is in pools and ditches, and the place which my father and grandfather as well as myself cultivated is now a wide strand.

What do you want to be done to remedy that? – It cannot be remedied. The Government of the Queen could not put it in order.

I suppose an embankment would be impossible to keep out the sea? – There is no relief in that way. It could be made, but it could not be kept up.

Has the planting of bent been tried as it was in the island here? It would be of no use. Where myself and my father and my grandfather tilled the land, some of it is covered ten feet deep by the sea. (Royal Commission Evidence 1883)$_{34}$.

The township of 'Hussaboste' is mentioned in the Charter of Inchaffrey dated 1389, but is reputed to have been washed away in the fifteenth century, and is believed to have been located off Baleshare, represented today only by the reef of Sgeir Husabost$_5$. The existence of this Charter and the fact that Husabost does not appear in documents later than the fifteenth century gives credence to the strong local tradition that the island of Baleshare (from *Baile Sear*, meaning 'East Town') formerly extended westwards, and the former existence of a *Baile Siar* or 'West Town' is implied. Sgeir a Chloidhean, west of Baleshare, is believed to represent an old flood gate [see quotation from Beveridge in respect of Monachs, below]. Reid's map of 1799$_{31}$, 43 years after the storm of 1756, shows Eachkamish joined to the remainder of Baleshare, though only by a narrow isthmus; this is now over 1km wide and apparently stable. It is interesting to note, however, that Walker (see above) describes Eachkamish as having been separated from North Uist rather than Baleshare, though this is almost certainly an error on his part. Professor Ritchie remarks that the parallel depressions in the machair between the arcs of old 'hooks' seem to have been flooded at one time from the east, not the west and, though he regards these as old features, he does not go so far as to suggest that they date from the mid-18th century[31].

MONACH ISLES (HEISKEIR)

Dr Alexander Carmichael ... takes Heisker to represent *Aoi-sgeir* or 'isthmus-skerry' although 'partly through the gradual subsidence of the land, and partly owing to the gradual dislodgement of the friable soil forming the isthmus, the isthmus by degree gave way to fords, and the fords broadened into a strait four and a half miles wide and four fathoms deep. Tradition still mentions the names of those who crossed these fords last, and the names of persons drowned in crossing'. The same writer adds (c. 1884) 'I know men who ploughed and reaped fields now under the sea' though this latter statement may apply to Harris instead of North Uist. (Beveridge 1911)[5].

[Carmichael] (*The Scottish Geographical Magazine*, vol.ii,pp.461-474) cites local tradition to the effect that Heisker was at one time joined both to North Uist and Benbecula; further adding – "Intelligent crofters informed the writer of having seen fragments of iron bolts sunk into certain low-water rocks in the Atlantic between Benbecula and h'Eisgeir [Heiskeir, Monachs]. These rocks lie at a distance of some miles from the land on either side, and are believed to have been the sites of embankments ... It is interesting and curious to find various submerged sites over the now wide and open sea still called by their place-names, as Sgeir a Chloidhean ('the barrier rock,' the site of a flood gate), Ceardach Ruadh ('the Red Smithy') and others." But Ceardach Ruadh still exists as a sandhill site on Baleshare island, immediately above its west shore and close to Sgeir Husabost. In this connection may be noticed the existence of a very small tidal rock off the west shore of North Uist, close to Dig Mhor, at the mouth of Loch Paible, and still known as *Airidh Nighean Ailein*, or 'the shieling of Allan's daughters. This is locally said to be an ancient pasturage, and the inference remains that when it acquired the name the coastline extended distinctly further into the ocean. (Beveridge 1911)[5].

PABBAY, SOUND OF HARRIS

In the great sandstorm of 1697 there was considerable damage done in South Pabbay by blown sand so that the farm of Middleton disappeared from the rent rolls. (Morrison 1967,47)[26].

There are about 300 Acres of what was formerly the best Arable and pasture Land in the Island of Pabbay, that are at present overwhelmed with sand. As the Sand blows from the Shore, the Sea advances, and accordingly upon the South West side of this Island, the Sea flows for a great Space, where many People still alive have reaped Crops of Grain. (Walker 1764,54)[36].

The south-east portion of Pabbay, according to tradition, formerly presented a large plain, consisting of sandy soil mixed with earth. So fertile was this plain that it obtained for the island the name of the Granary of Harris. It has all been swept away, and in its place a wide expanse of loose sand is seen, in some places leaving bare the subjacent rocks and beds of clay. The surface has diminished in level, but to what extent cannot be ascertained. Great banks of sand have, in some places, formed at the land edge of this plain, but there are no hillocks along the coast. ... Now, it is obvious that by far the greater portion of the sand has been swept into the sea; for, had it been scattered over the island, there was enough of it to cover its whole surface to a depth of several inches. (Macgillivray 1828)[21].

To the west of the landing place at Haltosh Point sand covers some largish patches near the coast, but never dominates the scene. This sand invasion is not recent, nor do the dunes appear to be spreading inland at the present day. ... If [the OSA] account is accurate, we may conclude that the sand invasion occurred before the end of the eighteenth century. Several facts suggest that there was little or no advance in the nineteenth. For one thing, the ruined walls of the largest deserted village, Baile-Lingay, are high upon on the sand-covered slopes, but have not been buried since 1842 by sand, nor has the smaller hamlet of Baile-fo-thuath. Again, Donald Maclean emphatically asserted that the dunes gave no trouble in the old days (i.e. the few years before 1842). Mr McGaskell of Rodel remembers that the seashore could be seen from the present croft when he was a boy, 40 years since. You cannot see it now, because the seaward dunes have grown in height. On the other hand, neither the croft, nor the ancient graveyard near it, have been overwhelmed by sand. The explanation seems to be a stabilization of the higher inland slopes of sand-covered rock into a closed turf that is the favourite pasture of the deer and sheep and cattle.... It seems likely that the change from much arable farming to pastoral grazing, lasting 90 years, has brought this stabilization about. (Elton 1938)[12].

The Bald map of Harris of 1805[3] shows the coastline much as it is today, yet an unpublished admiralty chart of 1857 in the National Maps Library in Edinburgh shows an extensive tidal inlet on the south side of Pabbay (Fig. 5.1). The coastline had been closed by the time of the survey for the first Ordnance Survey map of 1881, though a low-lying area is indicated just behind High Water Mark around the route of the Lingay Burn (Fig. 5.2). The area involved floods regularly in winter, and there may have been a particurly large body of water impounded behind a temporary dune for many months, which left a bare area

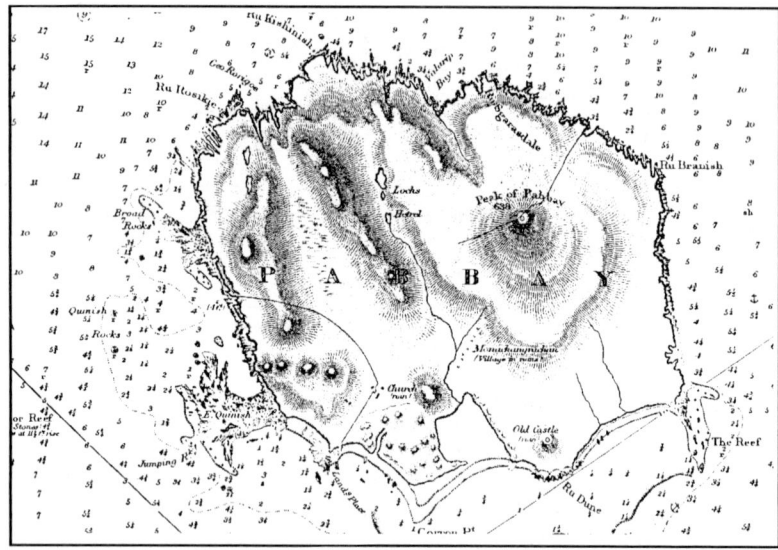

Figure 5.1. The coastline of Pabbay on Admiralty Chart 2642 of 1857, showing an inlet where there are dunes today. Reproduced by permission of the Trustees of the National Library of Scotland.

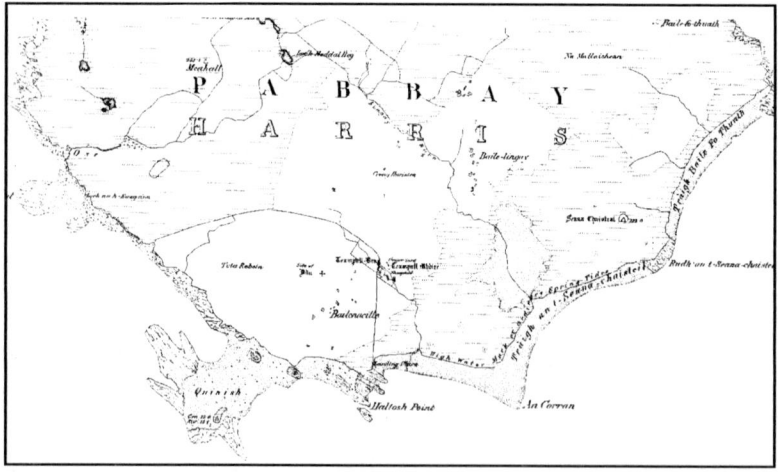

Figure 5.2. The first edition Ordnance Survey map of Pabbay (surveyed 1878, published 1881), showing the southern part of the island, with the dunes once again enclosing the inlet shown in Fig. 5.1. Reproduced by permission of the Trustees of the National Library of Scotland.

of sand when the barrier was eventually breached by the weight of water, prior to the date of the survey.

BERNERAY, SOUND OF HARRIS

Like Pabbay, [Berneray] suffered severely in the great sand storm of 1697, when the pendicle of Shiaby was covered by blown sand to a depth of several feet. (Morrison 1967,48)[26].

Shiabay remained part of Sir Norman's property and became more firmly a part of the island of Berneray when the great sandstorms of 1697 filled up the shallow sea between them. ... [Mary Macleod] lived with her mother in Shiabay in 1651 and remembered the area when it was all corn and pasture land and no part of it overrun with sand. As a young girl she used to herd calves there. ... Due perhaps to over cultivation, a long period of dry weather and the presence of gale force winds, the sandy soil of South Pabbay and North Berneray was driven to the south-west and covered the district of Shiabay to a depth in places of some sixty feet. The sub-tenants in the area managed to escape and through the good offices of Sir Norman Macleod were accommodated amicably on his other farms on the island. (Morrison 1982)[27].

There are about 300 Acres of the best Land in the Island of Bernera, entirely blown up with Sand in the same manner [as Pabbay], and the Drift has encroached so much upon Loch Bruist, a fresh Water Lake in the Island, that it is now firm Ground, where there was formerly a great Depth of Water interspersed with Islands. The Sand Drift is continuing to make great Devastation in the same way, along the west Coast of the main Land of Harris, and in all the lesser Islands which are adjacent. (Walker 1764, 54)[36].

HARRIS

...in the 1780s. Blown sand continued to play havoc on the estate in Berneray, Pabbay and Luskintyre. Ease of rents to the extent of £20 annually were given to the farm of Lingay, Pabbay; the rents in Berneray and Luskintyre were also reduced. (Morrison 1967,59)[26].

In several parts of the Harris, the Sand Drift from the Sea Shore, has made great Encroachments upon the Land. (Walker 1764,54)[36].

Of all these lands [Sound islands, Hushinish, Taransay] the soil is generally sandy; that especially which lies nearest the shore, from whence the sand is perpetually drifted by high winds, is pure shell, ground very fine. (OSA 57)[30].

... the prodigious variety of ruins of castles, fortresses, and villages, which, either raised on rocks and lofty eminences, have proudly defied the rage of the elements, and withstood the ravages of time, or which, built in lower situations, have been surrounded and overflowed by fresh water lakes, or arms of the sea bursting the slender barrier of a sandy or earthen bank, and jutting in through the land; besides the daily discovery of buildings on the western shore of most of these islands, exposed in consequence of the wasting of sand banks perpetually drifted by high winds (footnote, OSA,68-69)[30].

The tufts in which [marram] has been planted in Harris, allow the sand to drift between them to a certain extent, but they gradually enlarge. ... For [slopes], Mr Alexander Macleod found a most efficient preservative in square pieces of turf, cut from solid sward, and laid upon the drifting surface (Macgillivray 1828)[21].

'Mr Alexander Macleod' is undoubtedly 'An Dotair Ban', referred to in the next Chapter. In his wonderful book on Taransay, Bill Lawson[17] described how the Uidh, the isthmus in the centre of the island, was breached by the sea, probably around the year 1800. Campbell, the proprietor, blocked the breach with timber, planted marram, and restricted grazing, thus preventing the sea from turning Taransay into two islands. Bill points out that the place-names at Paible and local legends also suggest coastal change: Missader (erroneously located by the Ordnance Survey some 100m SW of its true position) is from the same root as Shader, *saetr*, an outfield, but is now a bay. Alexander Carmichael was shown human bones washed out of the sand by a Taransay woman, Mor Nic Cuien (Marion MacQueen), who also told him that the old folk claimed to have seen houses "under what is now neap high water, the land having been washed away".

GAELIC SOURCES

Gaelic poems and songs from the Outer Hebrides take us back rather further. The earliest reference in this field is a song by John MacCodrum, Bard to Sir James Macdonald of Sleat, called Smeorach Chlann Domhnaill (The Mavis of Clan Donald):

'S i 'n tir sgiamhach tir a' mhachair,
Tir nan dithean miogach daithte,
An tir laireach aigeach mhartach,
Tir an aigh gu brath nach gaisear

'N tir as boidhche ta ri fhaicinn,
'M bi fir og an comhdach dreachail,
Pailt na 's leor le por a' mhachair,
Spreidh air mointich, or air chlachan.

('Tis a beautiful land, the land of the plain,
the land of the smiling coloured flowers,
the land of mares and stallions and kine,
the land of good fortune which shall never be blighted.

The land most lovely to be seen,
where there are young men in comely apparel,
full plentiful in produce of the plain,
flocks on the moor, gold on the stones.)

As the bard himself mentions in the verse following those cited above, he was born (probably in 1693) at Aird an Runair in North Uist, now part of the RSPB's reserve at Balranald. The song quoted here was probably composed in the 1750s, when MacCodrum was living at Paible (Matheson 1938)[25].

MacCodrum's contemporary Alexander Macdonald (Alasdair Mhic Mhaighstir Alasdair) wrote Smeorach Chlann Raghnaill (The Mavis of Clan Ranald), which contains the following lines (Macdonald & Macdonald 1924)[18]:

'S a' Chreig-ghuirm a thogadh mise
An sgìr Chaisteil Bhuirgh nan cliar:
Tir tha daonnan ag cur thairis
Le tuil bhainne, mheala 's fhiona.

Sliochd nan eun bho 'n Chaisteal-thioram
'S bho Eilean-Fhionain nan gallan
Moch is feasgar togar m' iolach,
Seinn gu bileach, milis, mealach.

(I was raised in Creag Ghorm
In the parish of Borve Castle of the poets-bands
A land that constantly overflows
With flood of milk, of honey and of wine.
Offspring of the birds from Castle Tirrim
And from Islandfinnan of the saplings,
Morning and evening my cry is raised,
Singing billed, sweet and honeylike.)

While machair is not specifically mentioned, the lines clearly refer to the machair of Benbecula[6]. The poem expresses similar sentiments to those of MacCodrum about a fertile machair, and it is interesting to contrast these verses with contemporary accounts of the machair written (in English) by visitors, who unanimously agree that the west coast of the Uists at this time was virtually a desert of blown sand[2]. Macdonald is known to have visited MacCodrum at Paible around 1755, and both have been been accused of plagiarising the other[25] which, if true, would detract from their combined value in refuting the English accounts of widespread erosion.

Campbell and Collinson[10] give several South Uist songs (my italics)

XXIV. CHA DIRICH MI AN T-UCHD LE FONN
(I'll not climb the brae with song)

Nach dian seasamh air an fhonn	Who would never stand on soil
No air a' mhachaire lom,	Or upon the machair *bare*
Gun a bhotann bhith fo bhonn	Without shoes beneath his feet

XXVI. GURA MISE THA FO MHULAD AIR AN TULAICH LUIM
FHUAIR (Truly I am filled with sorrow)

Cha dirich mi bruthach	I can not climb the braeside
'S cha siubhail mi cluain	I can not pace the dell
Cha choisich mi 'n t-achadh	I can not walk the *firm* machair
No'm machaire cruaidh	Or step through the fields

XXXVI. ALEIN, ALEIN, 'S' FAD' AN CADAL
(Allan, Allan, long thou'rt sleeping)

Tir an eorna, tir na machrach	Land of barley, land of machair
Tir nan cruachan, tir nan daisean	Land of cornstacks, land of haymeadows
Tir nan sguaban, tir nan adag	Land of cornsheaves, land of corn-stooks
Far am faighte 'm biadh gun airce	*Where food can be had in plenty.*

Clearly there seems to be a difference of perception between the visitors and the residents: the visiting writers almost universally describe a waste of drifting sand, while the contemporary poets extol the fertility of the machair. It is possible that the poets did not regard the sandy wastes as part of the machair – there were almost certainly fertile parts at all times, yet it is strange that there is no reference to drifting sands which can be traced in the Gaelic record – and, of course, we should not forget 'poetic licence', not to mention the limited Gaelic knowledge of the writer.

THE KELP INDUSTRY

Machair sand is deficient in certain minerals and, being very loose when dry, is susceptible to wind-blow. As chance would have it, the answer to these problems was close at hand, in the form of the huge quantities of tangle cast up on the adjacent beaches, torn from the vast kelp forests off the west coast by winter gales (Fig. 5.3). This has long been spread on the surface of the machair and, though the bulk seems almost to disappear, the nutrients find their way into the soils,

Stewart Angus, February 1993

Figure 5.3. Tangle washed up on the beach, Howmore, South Uist

while the glutinous 'alginates' bind the sand. As described above by Macdonald, seaweed was used successfully to reclaim a large area of land at Paible in North Uist[19].

Kelp manufacture was begun in earnest in 1735, and provided a valuable source of income until the market collapsed in 1822[5,16,2]. Though 'kelp' is often used as an alternative name for tangle *Laminaria*, strictly speaking kelp is the burnt product of tangle; the kelp was used in the manufacture of soap, glass, and to obtain potash, sodium and iodine. Assisted by this source of income and the availability of potatoes (introduced in 1743), the population of the Outer Hebrides grew dramatically, increasing from 13,623 in 1755 to 35,563 in 1841. The contemporary accounts are virtually unanimous in condemning the neglect of the land due to a preoccupation with kelp, and some writers (e.g. the Old Statistical Account for Harris[30]) have suggested that the seaweed which would have been spread on the land to fertilise and stabilise the sand was instead burnt as kelp, to the extent that the landlord's Factor in Lewis was asked to punish anyone found using seaweed for any purpose of his own[16]. Macdonald[19] noted that the land was required only to provide the kelp producers with "some milk, a few carcases of lean sheep, horses, or cattle, and a wretched crop of barley, black oats and potatoes". Kelp was made from storm-cast tangle or *feamain dhearg* on the west coast, but on the east coast and on the islands in the Sound of Harris from growing seaweeds – probably wracks – *feamain dubh*[27], which had to be cut.

The detrimental effects of the kelp boom and the high human population left in its wake were not confined to a reduction in seaweed application. Horse and foot traffic increased; fallow periods had shortened so that cultivation became almost continuous in some areas due to the increase in population, and marram became substituted for straw as a thatching material due to a shortage of grain crops, while rotations for marram cropping were shortened[24,32,2].

About twenty tons of wet seaweed was used to make one ton of kelp. The price began to rise after Britain entered the Seven Years War with France in 1756, and reached a maximum of £22 per ton during the American War of Independence, of which the crofter got less than 10%. This raised huge profits for the landowners, to the extent that the island of Berneray, Harris, produced up to £2,400 revenue a year[27].

Though local people probably gained some additional income from kelp burning, vast profits were made by the landlords who owned the seaweed rights, and they developed a lifestyle in accordance with this income. When the boom ended, the income dried up suddenly, and the human population was, in their view, higher than the land could support. Sheep became their only alternative source of income, and the landlords evicted the human population to make way for the new livestock. Thus the kelp boom laid the foundations for the Clearances.

LAND IMPROVEMENT

Many of the sources quoted above make reference to measures taken to stabilise the blown sand. The importance of marram came to be recognised by proprietors, who introduced leases requiring any breaches in the vegetation surface to be plugged with marram, while the digging up of Lady's Bedstraw for use in dyeing was prohibited. The *New Statistical Account* for South Uist[29], quoted above, gives considerable detail on the measures employed. Marram planting was widespread, and local tradition holds that it was transported by boats on the network of inland lochs. It is certain that these lochs were once used for transport, as by the end of the eighteenth century there were concerns that the infilling of lochs with blown sand was threatening the use of the lochs as navigable links[1].

In addition to the sand stabilisation, there was a lengthy programme of drainage, which substantially changed the pattern of inland water in the Uists. The relevance of the loch drainage to erosion is that the water table would have been lowered over a wide area, drying out sand which would not hitherto have been liable to sand-blow. It is quite possible that the scale of sand drift in the

17th–18th centuries was linked to this drainage as well as to the kelp boom and to a lack of awareness of the measures which could be taken to prevent sand blow in connection with machair cultivation.

It is generally agreed that the worst of the machair erosion was over by the end of the last century, and that the problems of the twentieth century are of a more local nature[2]. These problems are examined in the next chapter.

6

MODERN MACHAIR EROSION

━━━►◄━━━

AIR MACHAIR BHARABHAIS	BARVAS MACHAIR
a' ghaoth	*the wind*
a'siabadh na gainmhich	*sweeping sand*
a'feannadh na machrach	*flays the machair*
ag ìsleachadh cnuic	*reduces the hills*
ar n-eachdraidh	*of our history*
gu rannsachail gu fàsach	*winnowing to desert*
rabaidean èasgaidh	*eager rabbits*
ag àiteach agus ag àrach	*cultivate and breed*
far na thogadh ar sinnsir	*where our ancestors were reared*
is far an deach an càradh	*and where they were interred*
an cinn-san an-diugh	*today their heads*
a' sgàineadh uachdar	*crack the surface*
mo sheallaidh	*of my vision*
aiseirigh na bochdainn	*miserable resurrection*
gun dùil rithe	*unforeseen*
Iain Moireach	John Murray[31]

Though there is a perception that machair erosion is increasing, this all depends on your point of reference in place and time. Machair is naturally subject to change, and erosion and deposition are not only natural, but essential, to machair systems. It could be argued, however, that if a problem is perceived by either local people or visitors, then a problem exists; human perception of the machair is arguably a vital component of the habitat, and one which should be taken very seriously. There is no doubt at all that a growing awareness on everyone's part can alleviate many problems. This chapter makes an examination of different machair systems, then goes on to address the various problems of machair erosion today.

The previous chapter demonstrated that large-scale change is an inherent feature of machair, and that seemingly intractable problems can be resolved.

Mather and Ritchie's detailed study[20] of 98 beach units in the Outer Hebrides revealed that 38% of these were 'truncated', with dunes absent or negligible. Dunes not only intercept much of the wind-blown sand from the beach, but afford the machair some protection from wave-induced erosion.

Though this situation may appear very serious, the international context gives a wider perspective. In 1987, the IGU Commission on the Coastal Environment analysed information from 127 countries, and concluded that over the last century, more than 70% of the world's sandy coastline had retreated, while only 10% had advanced[4]. There are parts of Cape Breton which lost between 700 and 800m width of coastline between 1881 and 1922, while parts of the Gulf of Audierne in Brittany lost 150m between 1952 and 1969[13].

As shown below, the problems of coastal erosion in the Western Isles may be exacerbated by an increasing relative sea level.

By definition, machairs are dynamic systems, and some change must be expected. Indeed, conservationists regard this dynamism as a positive and essential feature, but they also acknowledge the human perspective as essential to any evaluation (the positive role of people in machair habitat creation and maintenance is undisputed), as well as the possibility of loss of important habitats from machair systems.

A distinction should be drawn between wave-induced erosion of the coastal edge and the wind-related erosion which tends to affect only exposed machair surfaces (though wind obviously blows sand from the beach, this is not regarded as erosion).

During storm surges, there is not only violent wave action, but astronomical high tides (where the level is determined solely by the relative position of the Earth, Sun and Moon) may become even higher due to low barometric pressure, so that such conditions can cause severe damage to vulnerable coasts. The 38% of beaches which have no dune ridge are particularly vulnerable to undercutting during storm surges, and even the ridges of shingle on beaches may be of little protective value to the exposed, steep machair edge or 'front'. When coastal sand and even shingle are saturated by water, as they would be in violent storms, the sediments no longer absorb wave energy, but act instead as though they were impervious, so that high-energy waves can attack the coast[23]. Severe storm surges such as the 'Braer' storm of 4 January 1993, named after the tanker driven ashore in Shetland, may bring about rapid and significant change locally. Often some of the lost sand is redeposited the following summer, sometimes the process may take several years, and occasionally the sand may lie offshore for much longer. Likewise sand which is blown from exposed parts of the machair is redeposited downwind.

On these 'truncated' coasts with eroding machair fronts, the last sections to go are usually the upper layers of vegetation and root systems, which overhang the undercut sand, ultimately slumping downwards and breaking off under their own weight. This demonstrates the importance of the vegetation surface in maintaining the machair edge, and overgrazing, or any damage to the vegetation surface such as sand extraction or burrowing may accelerate the process of coastal retreat. The fact that machair slopes downwards away from the sea sometimes contributes to its downfall, as drivers who feel compelled to remain in their cars have to drive right to the machair front to obtain a sea view, where the weight of their vehicles further destabilises the edge.

Though the coastal edge may advance landwards, and give the impression that the sand is 'lost', it is merely redistributed: all erosion must ultimately result in deposition, and the sand may be blown inland, driven along the coast, or be washed offshore. Wherever it lies, it remains within the 'system', where it serves as a potential source of sand for future deposition on the land.

The winds which contributed to the formation and maintenance of the machair also threaten its stability. Chapter 7 (Fig. 7.4) shows that average wind speeds in the Outer Hebrides are very high. Almost all of the machair coastline of the Western Isles is 'very exposed', with mean annual wind speeds of 6.2-8 m/s; only the west coast of South Harris, Tràigh Mheilein in North Harris, and Stornoway Airport are 'exposed', with speeds ranging from 4.4 to 6.2 m/s. Dry, exposed sand will begin to move at wind speeds of 4.44m/s[20],so that all of the main machair coasts are susceptible to windblow for more than half the year. As only dry sand grains which are unprotected by vegetation move readily, however, most intertidal sand and sand below the water table will be protected, and the 263 'rain days' (days with measurable rain) are fairly evenly spread over the year as are the 'wet days' (>1mm rain)[23]. The western Highlands often experience a period in late spring when there is little rainfall and there has been little vegetative growth; the consequent drought in the well-drained soils makes any exposed surfaces very vulnerable to any strong winds during or following this period[20]. In his PhD Thesis[24], William Ritchie pointed out that the period of erosion in the 17th century (see Chapter 5) followed the worst drought in the history of Scotland, in 1652.

It is possible to control wind-related erosion more easily than wave-induced erosion, merely by protecting the surface or, ideally, by preventing surface damage in the first place.

Sometimes, unfortunately, well-meaning solutions to problems can make the erosion worse, transfer it along the coast, or create new problems.

The role of water in controlling erosion should not be overlooked. As dry sand blows more easily, drainage of machair systems may have undesirable consequences unless well-planned. Conversely, standing water may induce or increase erosion problems by lapping at exposed edges and possibly inhibiting the growth of vegetation, while streams may remove large amounts of sediment

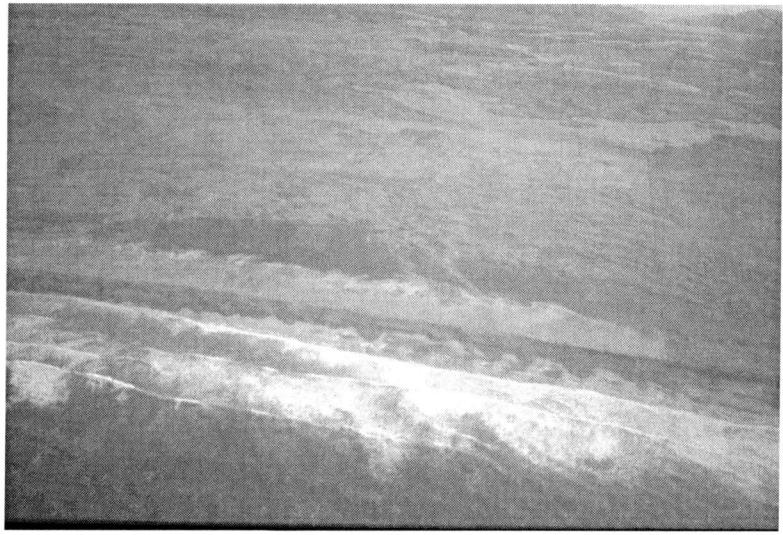

Stewart Angus, February 1993

Figure 6.1. Winter lochs, Ormiclete, South Uist. This standing water is seasonal, reflecting the winter rise in the water table.

to another part of the system. An understanding of the seasonal variation in water table is almost as essential to the study of machair erosion problems as the movement of the sediment (Fig. 6.1).

The section below reviews the current and recent situation in particular locations, beginning with the north of Lewis and working south.

LEWIS

One of the problems in any machair stabilisation project is locating a source of marram grass without causing additional damage. Tràigh Mhor, Tolsta, seems to be one of the few beaches which is accreting (building), though the system as a whole is probably vulnerable due to very high grazing pressure by sheep and rabbits. This is probably the only system in eastern Lewis where shingle is completely lacking.

The favourable situation at Tràigh Mhor is in marked contrast to that at neighbouring Garry, where erosion has been extreme. Visitor pressure at Garry has been high ever since Lord Leverhulme's road from Tolsta was surfaced in the late 1960s, and grazing pressure has also been high, even though the machair does not form a part of any common grazings – it has been grazed because of the lack of fencing. In 1986 The Countryside Commission for Scotland embarked on a small exclosure project, where part of the main machair was fenced off and cars were prevented from gaining access to a larger section of the machair. Around the same time, access to Tràigh Mhor was improved in the hope of diverting visitors there. Despite these measures, sheep continued to gain access to the fenced area, and vandalism of the fence around the car park meant that on one Sunday in May 1987, there were 85 cars parked on this very small machair. Despite these setbacks, Scottish Natural Heritage aided the owners, Stornoway Trust, to extend the fencing project, so that the whole machair is now protected, at least in theory, but rabbit damage continues, and sheep somehow seem to find their way into a well-fenced machair. While recovery continues, it is rather slower than it should be.

Further south, at Gress, the conglomerate cliff below Gress Lodge is eroding, and rubble has been dumped at the outflow of the river to prevent further erosion of the north machair (Fig. 6.2). The boulders are so large, however, that they are more likely to promote scour around their margins than solve any problems. There are also corridor blowouts in the machair which occasionally lead to sand being deposited on the main road, but these revegetate in summer with annual plants, and they are not serious erosion features. The northern machair seems to be established on a shingle base, probably formed from material washed out of the conglomerate cliffs, and has developed a series of hooks as it built, leading to the formation of saltmarsh inland, in a manner analogous to Tong[28]. The beach and machair are now the subject of a Coast Protection Order, and the northern section is a Site of Special Scientific Interest, so that there is an element of legal protection for the system. Ritchie and Mather have pointed out[28] that the removal of shingle would be even more damaging than the removal of sand from this system.

Coll has undergone considerable changes within my own lifetime. When I first went swimming there as a child in the 1960s, the beach was largely sandy, but within ten years the beach surface had been eroded down to a shingle substrate, especially around the estuary of the Allt an t-Sniomh, known locally as the 'Angus River'. Since then erosion has continued, and a short sea wall has been built under a section of conglomerate cliff, while extensive remodelling of the machair has taken place in the area where the road crosses the Allt an t-Sniomh. In August 1950, when Stornoway Trust was under considerable pressure from building contractors to open at least one of the systems near Stornoway to sand extraction, Trustees complained that men had to be sent out to Coll to clear the road of wind-blown sand, yet the Coll people had fenced off the machair to prevent anyone removing sand.

Stewart Angus, November 1996

Fingure 6.2. Boulders dumped on the estuary of the Gress River, Lewis

Stewart Angus, June 1992

Figure 6.3. The renovated south wall of the Branahuie isthmus

At the head of Broad Bay, the erosion now threatening the margins of the airport has been addressed by the erection of sections of porous plastic fencing in front of the dune slopes, which should intercept any sand being blown landwards from the beach.

Erosion of the Branahuie isthmus had become of concern to the Board of Trade as long ago as 1933[32]. A low sea wall was built on the Broad Bay side of the Branahuie isthmus in 1957, while the more robust wall on the south side was built in 1961, and strengthened in 1986 (Fig. 6.3). The beach on the Broad Bay side has a clockwise rotation, so that material is moved from the east end westwards (threatening the historic Ui Church), and a series of groynes was placed on the beach in an attempt to counter-act this, with limited success; groynes on the south side were similarly ineffective, as is evident from aerial photographs, which demonstrate little effect on the beach shape. In their 1970 report, Ritchie and Mather[28] criticised the decision to terminate the Broad Bay wall at a point almost south of Langa Sgeir Mhor, as scouring was taking place behind the end of the wall, and the wall was later extended west to the Melbost road end. As part of the 1986 work on the south side, piles were driven into the beach to change the beach profile. Locally it is believed that this has led to much more severe wave action on the road to the extent that it is more often closed on very high tides during southerly gales, and the works have certainly led to a transfer of erosion westwards, to the extent that a shingle rampart has been constructed between the sea wall and Holm. Apparently it is cheaper to rebuild this after every breach than construct an additional sea wall. This form of 'soft engineering' is a welcome change from the brutal sea wall approach, which usually transfers the problem along the coast instead of solving it; shingle is a superb material for absorbing wave energy, whereas concrete merely reflects the waves.

Tràigh Sanda, at Eoropie, is one of the main sand extraction sites in Lewis, and has been for some years (Fig. 6.4). Fortunately the sand is extracted from the south end of the beach, where sand is accumulating, and evidence from aerial photographs suggests that there is a plentiful sand supply offshore. Much of the machair surface has been eroded down to the water table, leaving isolated 'islets' of machair which may in fact be secondary deposition rather than remnants of the original surface. A temporary respite from sand extraction in 1996 has led to the accumulation of deep, new dunes on the southern machair, which have been colonised by marram.

There is serious machair erosion, leaving a very dissected surface, to the north and south of Swainbost Sands. I recall a visit to this site on a particularly cold day, and I was struck by the sheets of ice representing a spring line at the base of a sandy slope.

Stewart Angus, June 1991

Figure 6.4. Sand extraction, Eoropie

Barvas probably displays the most spectacular 'inland' machair erosion in Scotland – there is one site in Ireland which may be even worse. Barvas was chosen by the Nature Conservancy Council (now Scottish Natural Heritage) as one of two study sites for an investigation of machair erosion in 1988: the intention was to select a site with high human activity (Barvas) and compare it with another eroded site where the changes seemed to be largely natural (Luskentyre), in the hope that lessons could be learned which could then be applied to machair erosion problems throughout the Outer Hebrides. The study involved fieldwork, interviews with local residents, and comparison of the 1946, 1967 and 1987 aerial pictures. Clearly the extensive sand extraction from Barvas has had some effect, not only in terms of the sand removed, but also in the form of tracking damage from the heavy lorries. The area west of the cemetery had undergone considerable change since 1970, but the most damaged area was at Cnoc Mor, some 600-800m south of the cemetery, where an extensive area has been completely denuded not only of vegetation but also of its sand cover (Fig 6.5. Deposits of blown sand around the margins had been colonised by Butterbur *Petasites hybridus*. In addition to the sand extraction, damage is believed by local people to have been caused by rabbits and ploughing for potatoes and leaving the land bare to the winds[14]. During strong winds, spirals of sand can be seen rising from Cnoc Mor to considerable heights.

Stewart Angus, March 1988

Figure 6.5. Aerial view of machair erosion, Cnoc Mor, Barvas

There are people in Arnol who can recall the land extending over 30m from the present coastal edge out to particular rocks[16].

Dalbeg, with its superbly situated loch and scenic outlook to sea, is in some ways the victim of its setting. The remaining machair is small, and situated at the road end, almost inviting motorists to drive to the machair edge and have their picnics looking out to sea, where the wave action is often spectacular. When seven pilot whales were driven ashore there in April 1992, some of them still alive, huge numbers of visitors converged on the site, parking right up to the machair edge, so that the combined weight of their vehicles has probably knocked a few years off the life expectancy of the machair front. Since then the community has taken steps to prevent vehicle access on to the machair by placing very large boulders around the road end, and one hopes that they have not been too late.

Neighbouring Dalmore has a cemetery, and is thus entitled to coastal protection from the Local Authority, which takes the form of a pebble-filled wire cage behind stout wooden beams.

Bosta in Bernera is a popular recreation beach with local people, but has suffered greatly from machair erosion in recent years, with complete sections lost.

Stewart Angus, August 1994

Figure 6.6. Caravans, Traigh na Berie, Uig. This activity has been regulated since the photograph was taken.

Tràigh na Berie in Uig was until recently the site of an unregulated caravan site, with almost a hundred caravans there at peak times (Fig. 6.6). The caravans were blamed for much of the instability by the H.R. Wallingford report on coastal erosion[16], but the caravanning was brought under formal management in 1995-96, in association with controlled car parking and interpretation by the Coastal Access Programme, a partnership of the Western Isles Council, Scottish Natural Heritage and Western Isles Enterprise (aided by European funding), which should alleviate the problems.

As a Lewisman, I have to confess that my favourite beaches are all in Harris or North Uist, but if I had to choose a Lewis beach, it would be Ardroil (often wrongly referred to by tourists as 'Uig' because it lies at the head of Camas Uig). Erosion in this site is somewhat eclipsed by the shocking eyesore of nearby Carnish Quarry (by which I mean the site closer to Carnish village – the other quarry is better landscaped), but the inner part of the system at the Ardroil road end is experiencing some minor undercutting, and 'concrete sandbags' have been placed on the edge in the form of a wall, with the usual lack of success with such solid structures: they are being undercut. The Coastal Access Programme plans to improve parking and access facilities here in 1997, in such a way as to protect the fragile machair front from further land-based pressure, which should alleviate the effects of marine undercutting, and should also prevent the

Stewart Angus, April 1992

Figure 6.7. Blown sand, Carnish, Uig, Lewis

threatened build-up of caravans displaced from Tràigh na Berie, where recently introduced regulation of the former wild caravanning has imposed an upper limit on their numbers. The erosion from the east end of the system seems to be balanced by sand accumulation towards the west, where there are some fine climbing dunes and active growth of marram on the dune fronts.

There is a great deal of blown sand and bare surfaces at Carnish (Fig. 6.7), where there is a serious rabbit problem. The depth of blown sand is probably so great in places that underlying vegetation is smothered.

Mangersta shows what can happen to a machair system if erosion progresses to its conclusion: the sand has been removed from a huge area down to the water table or mineral layer, and there is little evidence that there is excess sand in the intertidal area to produce natural regeneration[35].

HARRIS

When I first started visiting Hushinish in the early 1970s, there was a huge machair blowout just beyond the turn-off to the pier, originating from extensive sand extraction – possibly in response to the revelation that this beach has the highest shell content of any sand in the Western Isles (82.9%) (Fig. 6.8). In 1977 the sandpit was filled in by Comhairle nan Eilean, and this seems to have been

Stewart Angus, July 1977

Figure 6.8. Hushinish, North Harris, showing the serious erosion, since rectified

one of the more successful reinstatement projects in the Western Isles. Even with the old pit infilled, the road to the pier is still often inches deep in blown sand, and aerial pictures taken in June 1994 show an 'apron' of blown sand stretching from the beach almost to the pier area. The road to Hushinish township is very close to the beach, from which it is separated by a steep slope.

Much of the machair of Tràigh Mheilein is dominated by marram *Ammophila arenaria*, and the even lines of marram growth suggest that parts of it, at least, were planted, though a natural growth form cannot be ruled out. Locally, the dune front is gentle, with a healthy growth of marram, but in places the slope to the sea is steep, with slumped vegetation, and the machair between the beach and Loch Cravadale is much dissected by erosion scars possibly linked to the high rabbit population.

Luskentyre Banks (Figs 6.9, 6.10) was the subject of the second study commissioned from Aberdeen University in 1988 by the Nature Conservancy Council. This work revealed that the system is highly dynamic, with considerable changes having occurred over the forty years studied (Fig. 6.11). Around the northern cemetery, there were two corridor blowouts, one of them very large, associated with a stream and the route to the beach; to the south-west, the machair front appears stable, but had actually retreated some 15-20m, especially at the western 'corner', with a build up of sand to the SE of this. The main

Stewart Angus, September 1984

Figure 6.9. Luskentyre Banks from the air

Stewart Angus, September 1990

Figure 6.10. Luskentyre Banks, showing bare areas on steep slopes

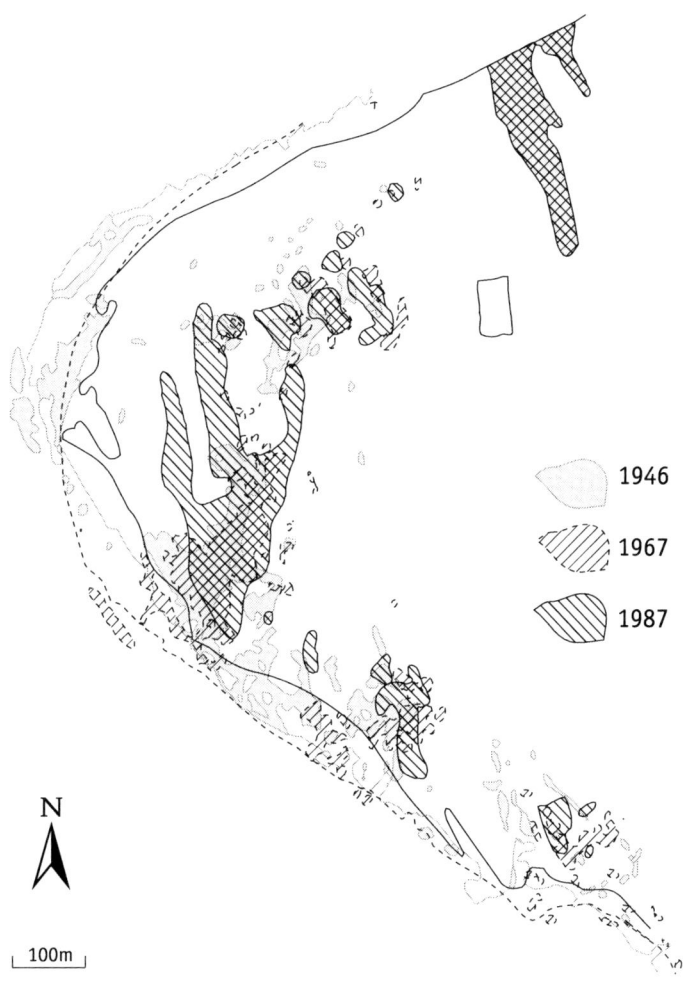

1946

1967

1987

N

100m

*Figure 6.11. Progress of erosion at Luskentyre Banks since 1946
(after Harris and Ritchie₁₅)*

blowout scars on the sandhills had migrated slightly to the NE and were much enlarged, with two N–S corridors developed after 1967; there had also been considerable, though smaller-scale, changes on the coast and machair between the high sandhills and the eastern cemetery. This cemetery is threatened by coastal erosion, and old cars dumped there in the past to combat the problem have become re-exposed, posing an additional problem. The researchers concluded that the system was in balance overall, but was clearly highly dynamic; no particular causes of erosion were identified[15,19].

Ritchie and Mather have pointed out that the Ordnance Survey map of 1901 shows the peninsula of Crago extending rather further northwards than it does today, but events of the last two decades or so suggest that the northern part of this dune system is very dynamic, with huge sections vanishing in the worst winter storms, and rebuilding over the following summer(s). It has also been pointed out that the saltmarsh soils fringing this inlet have the appearance of machair which has been converted to saltmarsh by marine inundation, which could make them very important in the study of the history of sea levels and coastal change in South Harris[28].

Tràigh Nisabost at Horgabost is part of the wider Luskentyre system and as one of the most sheltered beaches in Harris has high visitor pressure. It was a prime target for the Coastal Access Project, who funded the construction of a car park and surfacing of the access track with loose chippings to minimise erosion. Unfortunately this stimulated a planning application for a caravan site covering almost all of the machair which would seriously detract from the value of this site as an asset for the islands' increasingly important tourist industry. At the time of writing this proposal seems unlikely to proceed. The Horgabost machair was used for caravanning in the 1970s, but caravans were later excluded by the grazings committee following the Ritchie & Mather report[28], which warned that the situation was in the initial stages of the problems then being experienced at Achmelvich in Sutherland, where caravanning had been linked to what was then one of the worst cases of machair erosion in Scotland. The caravans have been creeping back to Horgabost in ever increasing numbers over the last few years. Meanwhile the new sand extraction sites on the headland are rather unsightly, but the Harris people have to obtain their sand somewhere, and it makes sense to take it from an area which was already damaged rather than open a new site.

There is an old sand extraction site north of the cemetery at Borvemore, which has revegetated well despite the steepness of the slopes. The ridge separating this hollow from the sea is disconcertingly narrow, and it could be breached by a severe storm. Just to the south, there is a rare case of pedestrian-induced erosion, where the easiest access from the cemetery car park to the beach is down the steep dune front, which now has a large and growing 'dent' in it due

Stewart Angus, May 1991

Figure 6.12. Erosion to machair front caused by pedestrian access, Borvemore, South Harris

to numerous pairs of feet pushing the vegetation and sand downwards (Fig. 6.12). This could be easily fixed by placing wire-linked fenceposts as a moveable 'staircase', and the site as a whole is targeted for action by the Coastal Access Programme. The machair edge is almost entirely steep with slumped sections of vegetation, and this system is actively eroding from the seaward edge.

At Borve Lodge, Lord Leverhulme is reported to have bulldozed a dune flat so that his view of the sea would be improved, probably shortly after he moved from Lewis to Harris in 1919. Westerly winds often blow sand on to the main road here, as the passage of the wind is uninterrupted.

The Scarasta system is extensive, and is for the most part in stable condition. Aerial photographs (Fig. 6.13) reveal a scalloped dune front, but the presence of a line of hummocky dunes, which seem to be growing, suggests that there is plentiful accretion locally (see also Chapter 4).

The south coast of the Northton isthmus is exposed to the shallow Sound of Harris, and a sea wall has been built at Tràigh na h-Uidhe to replace the old one, which was being severely undermined by wave action and there was a threat that the last two houses in the village could be cut off. North of Tràigh na Cleavag there is severe rabbit damage and erosion of a very low-lying area, to the extent that the sea could break through in a severe storm. The machair front is steep, with slumped sections of vegetation, and is badly broken by blowouts.

Stewart Angus, June 1993

Figure 6.13. Aerial view of Scarasta dunes

The problem has not been helped by vehicles, and the Northton township's closure of this system to unauthorised vehicles in 1996 was a step in the right direction. This site has been targeted by the Coastal Access Programme, so that convenient off-site parking should be available in the near future. The machair close to the old quarry at Chaipaval is severely dissected by erosion and the remaining vegetation very closely grazed.

The central isthmus of Taransay is heavily grazed, and the vegetation close-cropped even in summer, so it is hardly surprising that the machair is studded with blowouts, with three corridor blowouts that threaten to link the two beaches. There are cattle, sheep and deer on the island, but no rabbits, so that changes in management would alleviate the situation. At Corran Raah, the ridge leading out to the submerged isthmus (see Chapter 4) has very little vegetation – marram is virtually the only plant – and this system could be destabilised easily. At Paible, where the last residents – the Macraes – lived, the sea broke through the headland in the late 1970s and removed a portion of the machair, and this area seems to lose more ground with every year that passes[18]. The wreck of a catamaran in a blowout at Tràigh a'Siar is of some interest: the last occupant of this yacht was winched off safely after hitting trouble in a race … in the Caribbean![18]

SOUND OF HARRIS ISLANDS

Much of the machair surface on the west-facing system on Ensay has disappeared, and there is a large blowout to the west of Ensay House. There is an old burial ground on the western machair, where erosion often exposes human remains which have been studied by A.E.W. Miles[21]. Doubtless reduced grazing levels would alleviate the erosion problem.

Machair occupies a significant proportion of the island of Berneray in the Sound of Harris. Even the fairly recent 1:50,000 Ordnance Survey maps indicate a sand hill of some 20m elevation at Maol Bhan (NF909830), but this had ceased to exist by 1970, the victim of erosion[26]. The seaward margins of the machair are steep, and Ritchie has pointed out that such dune establishment as exists is due to redeposition of erosion products rather than to interception of incoming sediment. The machair close to the sea now seems stable, but is marked by numerous erosional scars; while it is tempting to attribute these scars to rabbit activity, it is said that the few rabbits which gained access to Berneray did not last long, and they have certainly been extinct there for many decades.

Neighbouring Boreray is perhaps my favourite area in all of the Outer Hebrides. The eastern beach enjoys a wonderfully picturesque setting, backed by a machair, abandoned settlement, and a machair loch impounded on the Atlantic side by a shingle ridge. Erosion is virtually absent, but the abandoned settlement, for all its visual attraction, is a reminder of how difficult life could be in these offshore islands when landlords failed to meet their obligations, and the loch – brackish because of input from the Atlantic – spread over the fields because of a lack of maintenance of the sluice, and a third of the arable area was lost[29]. Houses were rarely built on machair – the blackland was usually preferred – and there is something terribly poignant about the walls of old houses overgrown with marram.

While Berneray has its machair in the western half, Pabbay has its machair on the eastern portion of this conical island, due to the source of sand lying to the east in the shallow waters of the Sound of Harris. The history of the machair and its people has been ably chronicled by the prolific pen of Bill Lawson[17], with inputs from the equally indefatigable Professor William Ritchie[27]. As indicated in Chapter 5, there were significant changes in the sediment regime, notably during the storm of 1697, but the machair seems to have stabilised, and the area formerly occupied by an inlet now has a particularly rich flora[2]. The hummocky machair noted in Chapter 4 seems healthy (Fig. 4.26), and locally the dune front seems to be advancing, but there is quite a serious blowout problem on the higher parts of the machair inland, where the sward is very closely grazed.

NORTH UIST

Machairs Robach and Newton together provide one of the best locations for the study of machair formation, the former being one of the most dynamic systems in North Uist[25]. The line of sandhills just inland from the machair front on Lingay Strand is experiencing a moderate blowout problem (Fig. 6.14). The island of Oronsay was said by Ritchie[25] to have become detached from North Uist earlier this century, though 'Oronsay' suggests that it was a tidal island when the Norse named it.

Udal, an archaeological site which had human occupation for some 4,000 years, has been the subject of an extended excavation by Iain Crawford, who has not published many of his results, but such work as has appeared[8] has given important new insights into the interrelationships between people and machair over the millennia. Machair Leathann is a healthy system, with an abundant sand supply from the west and growing dunes[25]; locally there are steep slopes with vegetation slumping but much of the machair system has a gentle slope to the beach, often with a patchy but effective growth of colonising vegetation. Corran Aird a'Mhorain (Fig. 6.15) has accumulated in comparatively sheltered conditions, and takes the form of a very low-lying ridge of sand with a sparse growth of marram extending out to a sandhill at the tip.

Stewart Angus, August 1993

Figure 6.14. Lingay Strand, Newton, North Uist. The older dune ridge inland is showing signs of erosion, and is separated from the beach by a younger, lower machair.

Stewart Angus, June 1993

Figure 6.15. Corran Aird a'Mhorain, North Uist

It has been pointed out that the systems between Scolpaig and Hosta are usually based on shingle, at "no great depth beneath the sand"[26].

The beaches around Balranald are very scenic, and the setting of the villages adds considerably to their attraction. This area is not only a Site of Special Scientific Interest but is managed by the Royal Society for the Protection of Birds as a Nature Reserve, in close consultation with the local population. Coastal protection works, including a sea wall, were installed at Hougharry in 1992 to protect the road and reduce the amount of shingle washed over it in storms. The Aird an Runair beaches, notably Tràigh nam Faoghailean, form one of the highlights of the Reserve. Locally there is a problem with blowouts in the machair front, and the RSPB made attempts to counter this in the mid-1980s. Some of the steeper sections seem to be losing their battle with erosion, with the apparently imminent isolation of tall sections of machair front and the development of cauldron-like blowouts. H.R.Wallingford reported serious slumping of the machair front near Port Scolpaig, which they attributed to cattle trampling, and recommended fencing to exlude livestock from the machair edge[16]. Locally, blowouts have begun to revegetate naturally, though the effect may be seasonal (Fig. 6.16).

Vallay is a superb site, wonderfully demonstrating machair-saltmarsh and machair-sandflat transitions involving the slope of the machair plain. To the

Stewart Angus, August 1993

Figure 6.16. Vegetation recolonising erosion scar, Goular, Balranald

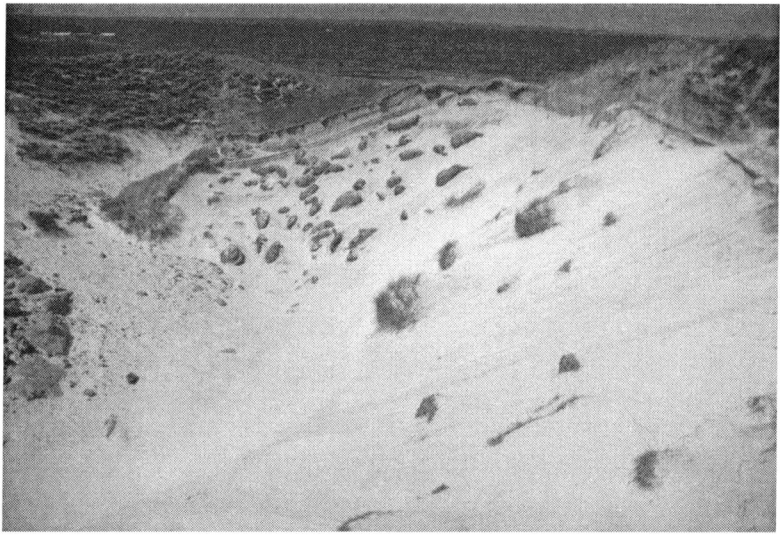

Stewart Angus, June 1992

Figure 6.17. Slumped vegetation in blowout, Vallay, North Uist

west, blowouts and sandhills prevail on higher machair (Fig. 6.17), especially to the west of Ceann Uachdarach, and this active system still conforms to Ritchie's description written in1968₂₅.

The lower sections of the coastline between Loch Paible and Kyles Paible are generally stable, but steeper sections are showing some signs of marine undercutting. It is interesting to note that on some steep sections where the machair front has collapsed the slumped vegetation seems to be promoting stabilisation at a lower level.

There is slumped fencing on the machair front at Ceardach Ruadh on Baleshare and rocks have been placed on and above the shore to give some protection from wave action. Large sections of the coastline are very low-lying, often with large areas of shingle on and above the beach, and these tend to be very stable.

The Monach Isles have few blowout scars, and these are largely restricted to the machair front, which has a few corridor blowouts. These islands form the largest Grey Seal breeding colony in Europe, and though a few seals utilise the corridor blowouts for access to the machair, for the most part they remain on the beach and cause no damage. The machair is generally in very good condition despite a high rabbit population.

BENBECULA

Though there is localised erosion of the dune face of An Tom, the beach around the airport, the system is substantially stable.

In response to a community request, Culla Bay is currently being developed by the Coastal Access Programme as the main visitor site for the Benbecula machair, though in time the old steadings at Nunton may be developed to provide a visitor centre, which could then be linked to access at Culla. A former sandpit here has been reinstated, but erosion is now exposing the infill material[22].

In early 1997, BBC *Reporting Scotland* reported erosion of the machair front at Lionacleit, to the extent that the school was regarded as threatened in the medium term. The coastline here is very steep in places, and there are all the signs of marine undercutting of the machair, while the growth of marram well inland suggests very active sandblow.

SOUTH UIST

The machair of South Uist is, in nature conservation terms, the best machair in the world[2]. At the north end, local erosion of the machair front at the Rangehead has led to reinstatement measures by the Army. Attempts to protect the coast in 1991 by the emplacement of rock-filled gabion (wire cage) baskets have succeeded at the north end (Fig. 6.18) but have largely failed in the south, despite

Stewart Angus, September 1993

Figure 6.18. Accumulation of sand on gabion baskets, West Gerinish, South Uist

similar methodology being applied. In the north, there seems to be a certain amount of mobile sand in the system, which has settled on the gently sloping gabions, whereas wave action has caused physical damage to the gabions in the south. The dumping of debris, including – at one stage – two Saracen armoured cars which were subsequently removed – has been for the most part unsuccessful, but the Army and local people are to be commended for recognition of their mistake and their attempts to rectify the situation by removing all the rubble and metallic sheeting which was placed in the blowouts.

The narrow dune ridge of Gualan is a dynamic structure. While a degree of change must be expected, erosion of parts of the ridge is giving rise to considerable local concern, not so much because of the the prospect of losing land, but because Gualan provides a great deal of shelter for the South Ford. The ridge contains huge amounts of shingle and, while there is undoubtedly a significant mobile element, in the medium term it seems unlikely to suffer more than occasional localised breaches, though a severe storm surge could conceivably cause more extensive damage.

The Drimsdale area is heavily infested with rabbits, but a current project initiated by the Uist staff of Scottish Natural Heritage may go some way towards solving the problem. They have identified an area delineated by deep ditches and machair lochs, and have worked with local people to provide rabbit fencing to isolate this area completely from the rest of South Uist. The idea is to eliminate (or at least control) rabbits within this area and, if the experiment is successful, extend the methods to other areas. There is also a problem here with marine undercutting of the machair front, as revealed by the fences now hanging in mid-air (Fig. 6.19), which the H.R. Wallingford study has blamed on localised shingle extraction[16].

The Howmore River is the only large river on the west coast of the Uists. Though rivers often rework the sediment around their mouth, the outlet of the Howmore River is relatively stable[35], despite coastal change to the north and south of the estuary (Fig. 6.20).

H.R. Wallingford noted that the beach level between Rubha Ardvule and Rubh Aird-mhicheil was very low, and attributed this to drawdown from successive storm surges. Blowouts have developed over much of the machair front, and rubbish has been dumped in some of these[16].

One of the most significant examples of machair erosion in recent times occurred at Bornish in the 1950s, when some 20ha of machair were reduced to bare sand, affecting adjacent areas with wind-blown sand, to the extent that one-third of the arable area of Bornish was affected in some way. The problem was blamed on injudicious machair cropping and rabbit damage, and in 1964, the Crofters Commission, the College of Agriculture and South Uist Estate enlisted

David Maclennan, February 1993

Figure 6.19. Hanging fenceline on retreating machair front, Drimsdale, South Uist

Stewart Angus, June 1990

Figure 6.20. Howmore River estuary, South Uist

the assistance of the Army, and levelled the remains of the blowout, then planted a suitable seed mix and manured the area with seaweed, dung and chemical fertiliser, succesfully reinstating the area$_{30}$.

BARRA

One of the two sand extraction sites in Barra was located at Eoligarry. When the operator applied for planning permission to continue with the work in 1993, his application was refused, as was his appeal, and Tangusdale is now the only site in Barra from which commercial sand extraction is permitted. The operator of the Tangusdale site, Hector Macneil, has carried out extensive reinstatement work on the Tangusdale machair for nearly twenty years, all voluntarily, and almost all at his own expense. While some of the methods, such as lines of barrels and buried scrap metal, could be improved upon, it has to be admitted that his efforts have largely succeeded.

The erosion of the Tràigh Eais system (Fig. 6.21) threatens to encroach not only on the Eoligarry machair but also on the road and even the airport, and the landowners, the Department of Agriculture and Fisheries for Scotland (now the Scottish Office Agriculture, Environment and Fisheries Department) had a

Stewart Angus, February 1993

Figure 6.21. Tràigh Eais, Eoligarry, Sout Uist

Stewart Angus, 1977

Figure 6.22. Geometric patterns of planted marram, Tràigh Eais, Barra

long-term marram planting programme, aimed at alleviating this problem, extending over more than twenty-five years. Sandblow fencing was also involved, giving a geometric pattern to the dunes (Fig. 6.22), while on steep slopes the marram was planted in a geotextile mat[3]. Though some progress was made, the work was hampered locally by the sheer volume of sand going through the blowout corridors but in general terms the project has been successful in stabilising the machair, but will not be continued[12]. The Eoligarry peninsula is unusual in NW Scotland in having 'through' blowout corridors, so that sand may blow from east to west as well as from west to east, though net transport of sand is in the latter direction.

The Tràigh Mhor in Barra has been used as the island airfield since 1936, but the operator of the air service wishes to replace the existing Twin Otter aircraft with larger machines, which they claim will ultimately require a fixed airstrip. A referendum held on Barra in 1996 voted firmly against the building of an airstrip on the Tràigh Mhor. A report commissioned by Scottish Natural Heritage from the University of Glasgow[12] warned that the introduction of an airstrip could have serious implications for sand exchange within the dunes and machair, as well as for the intertidal area, and even for the tidal currents in the Sound of Orosay. Thus the whole sediment regime of the system would be affected, probably promoting extensive scour within the Sound, while sand-laden winds from the west could affect aircraft stability. The southern erosion scars have been described as "some of the most spectacular erosion features in Britain"[12]. Grazing, particularly by the prolific rabbit population, has been a serious problem on this site, and not only threatens the existing machair, but also detracts from the effectiveness of the reinstatement measures.

There is some erosion of the small Sheader machair at the west end of Sandray, but the larger system which runs right across the eastern end of the island is experiencing quite serious erosion. The latter site has been called the 'wine glass' because of its appearance from the sea[11].

THE EROSION PROBLEM

If there is one conclusion which could be drawn from the studies on Barvas and Luskentyre commissioned from Aberdeen University, it is that each site is unique, and requires its own individual management policy. In an attempt to take a wider look at the problem, a colleague, Mary Elliott, and I reviewed the history and contemporary machair erosion problem in 1992[3].

Though rabbits were introduced to an island in West Loch Roag prior to 1549, they probably did not reach the Uists and Barra till the early 19th century and the first successful introduction to mainland Lewis and Harris may have been as late as 1865[1]. Organised rabbit control was practised as long ago as the 1890s in Barra[11], though it may not have been until the 1950s that rabbits became a serious problem as agricultural pests and as an important agent of erosion on the machair. Myxomatosis was then introduced to try and reduce the numbers; before this they had been controlled by trapping for food, but this ceased after the introduction of myxomatosis. Though the rabbit has recovered, the appetite of the public for their meat has not, and rabbits now probably present the greatest single problem to machair stability. They are a particular problem at Garry and Barvas (Lewis) and in parts of North and South Uist, and they are often most prevalent where there are other notable agents of erosion, as at Garry and Barvas. As previously mentioned, they are also very numerous at Eoligarry and around Drimsdale in South Uist. The threat comes not only from the burrowing, which is at its worst in major warrens (which are fortunately the easiest populations to control) but also from scrapes, which expose sand surfaces to wind action, and from their grazing.

Cattle dung is used for stabilising bare areas, particularly in the Uists (cattle have declined dramatically in Lewis and Harris), and includes a rich seed source. Sheep numbers have remained high, and there can be no denying that some systems are overgrazed by sheep as well as rabbits. Sheep are an additional problem in that they shelter in blowouts, thus enlarging them; cattle also shelter in blowouts, but at least their dung aids revegetation.

Grazing throughout the year is a problem on some systems from livestock and on all but a few of the offshore island systems with rabbits. The most sensitive time is late summer and autumn, when plants can be cropped before they have had a chance to set seed, eventually reducing the number of species – especially annual plants – and encouraging the spread of mosses[5]. Too little grazing, on the other hand, severely reduces the nature conservation interest, as the vegetation becomes choked and reduced in variety by a rank growth of Red Fescue *Festuca rubra*.

Commercial sand extraction is, at last, coming under effective planning control. Scottish Natural Heritage asked the Council to prepare a strategic policy to manage sand extraction, and SNH, WIIC and Western Isles Enterprise commissioned a report from the British Geological Survey (BGS)[7] which could form the basis of such a policy. At the time of writing the conclusions of this report are about to become incorporated in the Council's Minerals Strategy, which will form part of their Structure Plan, which the Council is obliged to follow in determining planning applications.

When the extension to No. 1 Pier in Stornoway was being built in 1948, the only suitable sand available near Stornoway was from Aignish, where extraction had been stopped because of coastal erosion. Before it was ascertained that sand from Carnish would comply with requirements, the Pier and Harbour

Commission shipped 250 tons of sand from Loch Sunart to avoid delaying the work[10].

The BGS report assessed the volume of sand on each island, and also looked at its characteristics in relation to the British Standards applied to the building industry. The overall conclusion of their very detailed work was that the foreseeable needs of the industry could be met from existing sites, and it would be better to continue at these rather than open up new locations[7].

In addition to building sand, there has also been considerable extraction for land improvement. The high shell content of machair sand makes it particularly suitable for neutralising acid peatlands, and large quantities (probably in excess of 100,000 tonnes) of shell sand were removed from beaches and machairs during the large-scale reseedings of the 1960s, mainly from Barvas and Eoropie with smaller amounts from Hushinish, Reef and Ardroil[28]. All but one of over 1,000 reseeding schemes assisted by the Western Isles Integrated Development Programme (IDP) involved shell sand[34]. Surprisingly few major erosion problems seem to have arisen as a result, possibly because people were sufficiently aware by this time to realise where the worst problems would arise; the people in Great Bernera, for example, refused to allow sand removal from Bosta because of its instability, so that it had to be imported from the uninhabited island of Little Bernera[33]. Many grazings committees now prohibit the removal of sand and shingle from the machair and foreshore, even for small-scale agricultural use, though this has traditionally been regarded as an inalienable right by crofters. There are instances of individuals obtaining their sand from the next village along the coast, but this is rarely tolerated for long. While it might seem preferable to remove sand from the foreshore rather than the dunes or machair, Mather and Ritchie have pointed out that this can be even more damaging, in that the sand supply for the active beach system is depleted, and add that "in some cases it is no exaggeration to say that it threatens the stability of the whole beach complex"[20].

Crofters have a right to extract sand without planning consent from dunes or machair "which they own or occupy, providing that the sand is for their own use and is for agricultural purposes". Though there is still a legal requirement to obtain the consent of the holder of the mineral rights (almost invariably retained by the landowner) the exemption from the planning legislation is often interpreted as a legal right to the sand, but it is most unlikely that anyone other than the landowner is going to insist on the full observation of the law in this respect. Having said this, much of the machair is of international importance for nature conservation, and consultation with Scottish Natural Heritage is usually required on designated areas before extraction may legally proceed, and the UK Government has a legal duty under the EC Habitats and Species Directive to safeguard the interest of Special Areas of Conservation. More and more grazings committees are becoming aware of their own role in preventing machair erosion – after all – crofters are the main victims of erosion, and my feeling is that in time this problem will be effectively controlled at the community level without the need for additional intervention or legislation. Studies of small-scale sand removal in Ireland[6] revealed that there was an adverse impact on tourism from vehicles actively involved in sand removal and from the abandoned machinery which seems to be the inevitable consequence of quarrying activity on any scale.

Shingle and/or boulder barriers protect the machair of the Uists (Fig. 6.23) and some of the systems in north-western Lewis. Some of those in Lewis are currently threatened by extraction proposals, with an intended market as ornamental garden stone. Most, if not all, of the coastal communities involved, as well as the local Planning Authority, are fortunately well aware of the potential for erosion if this is permitted and it is unlikely that any extraction will proceed. There are no safe local sources for such markets, other than the boulders within the exploited deposit at Carnish.

Marram *Ammophila arenaria* is the main stabilising plant of blown sand in the Outer Hebrides, and its value is widely appreciated by local people. It is still cropped occasionally for animal bedding, as has been noted at Barvas, and even for thatching material, but never on a scale which leads to problems. There is a strong tradition, especially in the Uists, that marram was introduced to the

Stewart Angus, August 1993

Figure 6.23. Shingle ridge on beach, Peninerine, South Uist

Outer Hebrides. This grass is native to Scotland, and it seems most unlikely that such a common and widespread species would be absent from such a suitable area, even allowing for the colonisation difficulties of islands. There is no doubt that marram was planted on a large scale, and it may be that some of the plants were imported from outside the Outer Hebrides.

Butterbur *Petasites hybridus* has become a problem in parts of Lewis: this invasive plant not only covers large areas of machair in spring and summer, thereby limiting its grazing value, but dies back more than other major machair plants in winter, exposing the soil surface to wind action[9].

Motor bikes and cars have become a problem at sites such as Garry, Gress and Reef in Lewis, and Northton in Harris, though, as described above, progress has been made in controlling vehicle access at most problem sites, thanks to growing awareness on the part of local residents. The heat of machair barbecues may kill vegetation (which might be invisible beneath the sand), creating or exacerbating erosion problems – the beach is the best place for a barbie! Tourism-derived erosion has increased recently in parts of Tiree in the Inner Hebrides because of windsurfers' vehicles on the turf and the cumulative effect of their habit of gaining access to beaches by jumping down steep, eroding machair fronts. Though the current low level of windsurfing in the Outer Hebrides tends to focus on robust sites, there is no reason why the sport should not expand to become the international attraction it is in Tiree, where the main feature is the availability of surfing whichever the wind direction – something the Western Isles could offer just as well.

Dumping of old vehicles, builders' rubble (Fig. 6.24), waste hay and silage has been a problem at some times in most parts of the Hebrides. It cannot be stressed too strongly that the dumping or placing of solid objects in eroded areas is likely to lead to further erosion by scouring rather than to any alleviation of the problem. While there are exceptions where dumping has worked, it is almost certain that more suitable materials such as porous fencing which 'break' the wind rather than deflect it, would have been more effective in these locations. There is the additional problem with any dumping that even if the materials become covered, history clearly demonstrates that these coasts are very dynamic indeed, and if scrap metal re-emerges, it does so in a dangerous, unsightly, rusting state, and is then very difficult to remove. It is interesting to note that the assessment of coastal sandpits carried out for the Minch Project concluded that to the casual observer disused sandpits had very little visual effect on the landscape, except where disused machinery was in evidence or buried fill was becoming re-exposed[22]. Given the increasing economic importance of tourism in the Western Isles such dumping should be stopped immediately, and consideration given to the removal of scrap which is still exposed, as on the island of Berneray in the Sound of Harris. Any vegetation or soil dumped to 'stabilise'

Stewart Angus, June 1991

Figure 6.24. Dumped spoil and rubble, Eoropie

machair fronts is not only unsightly but inevitably contains 'weeds' alien to the machair, and soil dumped at Tràigh na Berie in Uig has introduced nettles to a heavily used section of the dune front – scarcely suitable for a tourist attraction. Peat or seaweed are not only more successful stabilising agents, but they do not contain seeds of, or provide opportunities for, species which are aesthetically or ecologically undesirable. Regrading of slopes to reduce steep slopes, preferably to less than 35°, is essential in any reinstatement of sandpits[22]. The recommendations of the sandpit reinstatement study could also be applied to many blowouts, with the proviso that infill should *never* be used near the machair front.

As shown in Chapter 3, the coastline of the Outer Hebrides is slowly being submerged by a rising sea level, in addition to which there is a global rise in sea level. Recent work by the Permanent Service for Mean Sea Level, suggests that while the global average rise in sea level is of the order of 1–2mm/year, the rise in the Western Isles (based on records from Stornoway) has been tentatively calculated as 4.5±2.4mm/year[36], or a minimum of 24cm or maximum of 45cm over 100 years. Further work is required to ascertain just how serious this threat is likely to be in the islands, but even the most optimistic figure is worrying, and should provide much food for thought for those involved in any aspect of coastal management.

The previous chapter demonstrates that the historic record shows much more extensive erosion than is experienced today, though it is beyond doubt that in the lifetime of individuals stretches of coastline have been lost, and it is no comfort to a crofter who has just lost a few more metres of his land, that in time the situation will be in balance, or even that a crofter along the coast could be the beneficiary in the form of redeposited sediment.

The Local Authority is not obliged to become involved in coast protection unless property or services are threatened, and with the widespread scale of the problem, it is probably unreasonable to expect that the UK Government, or even the European Union, could justify extensive works in sparsely populated areas.

As identified above, each site requires its own solution to its own problems, but the first requirement of any major project must be an understanding of the coastal sediment budget. Beaches often occur in groups linked to an identifiable offshore area, and each of these self-contained sediment 'cells' requires its own study. The Western Isles has two cells – east and west – with the former divided into 4 sub-cells, and latter into 6[16].

Though the problem may be a growing one, I feel sure that the combination of increased academic interest with increased awareness of the causes of erosion within communities will, at the very least, lead to solutions on some of the more amenable sites. With a possible rise in storm surges linked to rising sea level, however, it may be a case of slowing the inevitable, but until the inevitable element is confirmed, we can but try.

7

CLIMATE

—————➤●◄—————

Benbecula is one of the most wind-smitten islands on a coast that holds the British record (the Shetlands are less windy than the northern Hebrides). Winter gales seem well-nigh continuous; even in summer life on that west coast is bedevilled by that unrelenting blow. Tourist literature describing beaches here as fabulous say no less than truth, but small use can be made of them. The most hardened crofters have described their life to me as 'like living on the deck of a ship'. But their Norse blood must still be strong, for they are not dismayed. 'Och,' said one, 'it iss not so bad. We haff some shelter under the lee of Rockall.' (Murray 1966)[35].

Even by British standards, the weather is a bit of an obsession in the Western Isles. In bright sunshine and light winds, island scenery can make the spirit soar, while a winter gale can quickly transform the sea into an impassable barrier which disrupts lives in ways that only islanders can truly understand. In communities which have long earned their living from the sea, shared opinions on the vagaries of the weather can make the difference between life and death.

Mean Temperature 1941-70 °C	Jan	Feb	Mar	Apr	May
Mean daily	4.1	4.1	5.6	6.9	9.1
Mean daily max temp	6.1	6.5	8.3	10.1	12.3
Mean daily min temp	2	1.6	2.9	3.8	6
Mean relative humidity (%)	88	86	85	82	80
Av monthly rainfall as % of year's fall	10	8	7	>6	<6
Av no. days measurable (0.2mm or more) rain	25	22	24	19	19
Av daily sunshine (hours)	1.2	2.3	3.5	5.1	6
Av days with snow (1941-70)	8	7	6	3	1
Days with air frost	10	10	7	5	
Days with gale at some time in day	9	7	4	3	1

Table 7.1. Climate statistics for the Western Isles

Local people are often good *màirnealaichean* – weather forecasters – and Raghnall Mac Ille Dhuibh of Edinburgh University has summarised some of the weather lore in a superb series of articles in the *West Highland Free Press* in 1986[27].

You can, if you wish, relate to the climate in terms of the standard facts and figures (Table 7.1) but scientists takes things a bit further than this. Biologists have long realised that plants and animals, like ourselves, respond to climate as a whole rather than to rainfall, wind speed or temperature individually. The Macaulay Institute for Soil Research (now the Macaulay Land Use Research Institute) has published a series of three excellent maps of bioclimate; these tend to employ terms such as "oceanic perhumid oroarctic" which are not exactly user-friendly. I have broken down these maps into the components such as wind speed and accumulated temperature (see below for definition) for the sake of comprehensibility, but the reader should note that the various components act in unison as far as the environment is concerned.

Climate also exerts an important influence on soils, and an introduction to soils is given in Chapter 8.

The Western Isles lie just to the south of one of the main routes followed by the numerous depressions or low-pressure areas which develop out in the North Atlantic. The weather we experience as these depressions meet our coasts depends on where the depression originated, how deep it is, and on which part of it passes over us, as well as its position in relation to other low or high pressure areas.

Jun	Jul	Aug	Sep	Oct	Nov	Dec	Year
11.7	12.9	12.9	11.6	9.5	6.4	4.9	8.3
14.8	15.7	15.8	14.4	12	8.8	7	11
8.5	10.1	10	8.8	6.9	4	2.9	5.6
81	84	85	85	87	87	88	85
6	7	8	10	10	10	12	1094mm
17	21	22	21	23	24	26	263
5.8	4.3	4.3	3.5	2.4	1.6	0.8	3.4
				0.5	3	6	35
				<1	6	8	47
1	1	1	3	6	6	8	50

Figures are for Stornoway (from Manley 1979)

Climate

High-pressure systems (anticyclones) also pass over the Western Isles, but they do so infrequently, and most of our weather is associated with depressions. Occasionally anticyclones sit over or near the islands for weeks on end, often in spring, but sometimes in winter.

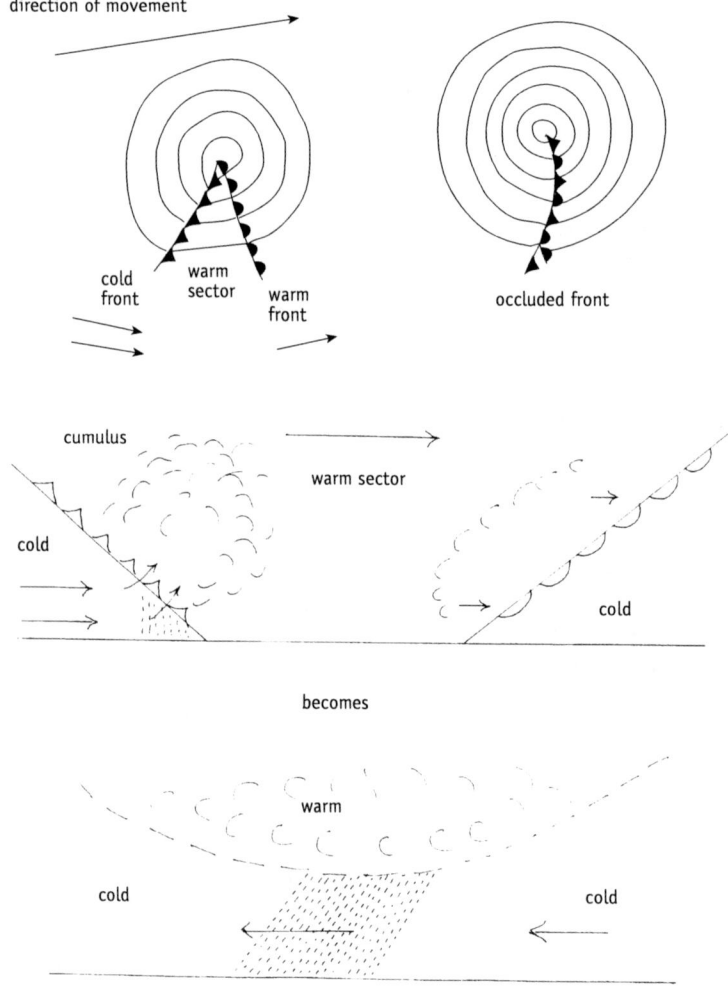

Figure 7.1. Warm, cold and occulded fronts

Depressions are formed when cold, dense, polar air meets warm, light subtropical air, and the interaction of these air masses is instrumental in determining the nature of the weather experienced on the land or sea below. The cold air compresses the warm air into a tongue-shaped mass which is bounded on its leading edge by a warm front and on its trailing edge by a cold front; both forms of fronts are familiar from newspaper and television diagrams. The cold front eventually catches up with the warm front, forcing the warm air to rise above the cold, so that the front is then said to be 'occluded' (Fig. 7.1). It is fascinating to watch the progress of these weather systems over a period of hours from the west coast or from the deck of a boat plying to or from the offshore islands.

Warm fronts are heralded by the feathery trails of cirrus, clouds associated with strong air streams at high altitudes, which are then replaced by extensive banks of stratus, the low, grey cloud which obscures the sky for so much of the time in the Western Isles. There may be a lengthy spell of rain, with particularly heavy precipitation as the front itself passes.

Once the warm front has moved through, though, there is usually a noticeable rise in the temperature as the warm sector of the depression is encountered. During the warmer months the warm sector can provide fine weather, but otherwise the high moisture content of the warm air may be sufficiently cooled by prolonged proximity to the sea to condense as stratus

Stewart Angus, June 1984

Figure 7.2. Low stratus clouds, South Lochs, Lewis

(Fig. 7.2) which may give rise to drizzle, and may even descend to sea level as fog. The term 'fog' is employed when visibility is less than 1000m or 1100yds[30].

An approaching cold front can be recognised by the line of towering cumulus clouds, formed by condensation as the warm air before the front is driven rapidly upwards. There is a sudden drop in the temperature as the front itself arrives, and the cumulus clouds give heavy rain, and sometimes even hail or thunder. Cold fronts move rapidly, and the downpour soon stops.

Many depressions are occluded by the time they reach the Western Isles. These carry heavy cloud and cause rain, but there is no change in temperature as they pass because the two ends of the cold front have met at ground level (Fig. 7.1). Occluded fronts move slowly and erratically, and predicting their behaviour, and that of the forthcoming weather, is very difficult[30].

High-pressure systems (anticyclones) move very slowly and may persist for weeks, though in some years they may not develop at all[30]. Although they do not have fronts, and often give prolonged periods of fine, calm weather with clear skies, they are sometimes associated with low stratus or fog.

Of course the above explanation of Hebridean weather is a simplified one, employing only 'text-book' examples of depressions. Whilst such situations occur, our weather charts are usually complicated by the presence of secondary features such as high-pressure ridges or low-pressure troughs.

Though the foregoing explanations of fronts may have been lost on islanders who lived in the days before radio weather forecasts or even barometers, Raghnall Mac Ille Dhuibh[27] revealed that the old boys knew a thing or two about their behaviour, for example: "*Gaoth 'n iar an déidh uisge reamhair* (west wind after thick rain); *Gaoth an iar gun fhrois, bidh i triall gu deas* (West wind without showers will back to the south.)".

The first systematic measurements of weather in the Western Isles were made at Lews Castle in 1856 and continued there until 1884. Meanwhile, Stornoway Coastguard had started to take measurements on behalf of the Meteorological Office in 1881. The weather station was transferred to Stornoway Airport in 1942, where it is still located. The noted climatologist Gordon Manley[31] observed that as the airport is rather more exposed than the Castle Grounds, and the Coastguard measurements were taken at different times of day and at more than one location, accurate historical comparisons of past records with those obtained since 1942 are not really possible.

Meteorological observations also date from 1942 at Balivanich Airport, and many records have been kept by lighthouse keepers, notably at the Butt of Lewis[31], a name eminently familiar from the Shipping Forecast's reports from Coastal Stations.

Stewart Angus, June 1993

Figure 7.3. Automatic weather buoy, North Rona

WIND

One of the most terrifying experiences of my life took place on the *bealach* between the Clisham and Tomnival in May, 1986, when I was twice lifted right off my feet by the wind, a wind so strong that we could not even contemplate a return trip downhill into the teeth of it and had to make a five-mile detour. The strongest gusts made a roaring sound as they approached, barely giving us time to crouch to the ground before they hit us, leaving us with an overwhelming sense of helplessness in the face of the awesome power of the gale. Water in streams was blown uphill, something that can quite often be seen on one of the waterfalls on Seaforth Island. When the sea and the wind act together, the insignificance of human endeavour becomes only too apparent, and even the most robust man-made coastal structures can sustain serious damage.

In the days when your house roof was thatch, not even the most intricate arrangement of ropes and netting could save you from the explorations of the strongest gusts, and most older houses were located and aligned so as to maximise shelter, incidentally creating a settlement pattern that seemed to sit very comfortably in the natural landscape, but now largely lost with the advent of stronger housing materials and a desire for a view overcoming that for shelter.

As most depressions pass to the north of us and their winds circulate in an anti-clockwise direction, it is not surprising that our prevailing winds are from the south and south-west (Fig. 7.1). A meteorological rule known as 'Buys Ballots Law' states that if you stand with your back to the wind in the northern hemisphere, the low pressure area must be to your left. Winds associated with northern hemisphere high-pressure systems circulate clockwise. Changes in wind direction, for example those associated with the different sectors of a depression, are denoted by 'backing' if the change is anticlockwise (in respect of the compass), and 'veering' if the change is clockwise.

In an average year 50 days with gale-force winds are recorded at Stornoway[31] and the average wind speed over the year is 14.4 knots (7.4m/s)[9]. Indeed, it has been said that the north-west of Scotland has a higher sustained wind speed than any other inhabited part of the world[16].

Most gales occur during the winter months (Table 7.1) when the difference in temperature between the polar and subtropical airstreams of the North Atlantic tends to create much deeper depressions, which in turn give rise to stronger winds.

The distribution of average wind speeds in Scotland has been plotted by the Macaulay Institute for Soil Research (Fig. 7.4). Only a few long, deep valleys were judged to be 'moderately exposed' with average wind speeds below 8.5 knots. Most of the land area of the islands is 'very exposed' with average wind speeds of 12–15.5 knots, but only the summit areas of the highest hills are 'extremely exposed'.

High average wind speeds tend to stunt the upward growth of plants and encourage dwarf forms, e.g. of Heather *Calluna vulgaris* and Juniper *Juniperus communis*. The prostrate form of Heather recorded from St Kilda may be linked more to salt-burn than to altitudinal descent of the mountain growth form[33]. Wind also has a profound effect on plant growth in the Western Isles in that it is salt-laden, particularly when the wind has come from the west or south-west (as it usually does), having passed over long stretches of wave-torn ocean. As would be expected, the salt content of the air drops off significantly with increasing distance from the sea, though local variations in topography affect this relationship. When the maximum gust speed exceeds 30 knots the salt content of the wind increases proportionately, as there is more sea spray – and the stronger winds carry the spray further inland[22,40]. Studies on the Monach Isles National Nature Reserve suggest that salt-laden winds affect taller plants most, so that exposed coastal plant communities tend to consist mainly of low-growth plant forms – not just because of the stunting effect of the wind, but also because of the salt it carries[40]. Salt-laden winds will also tend to increase the salinity of lochs close to the west coast.

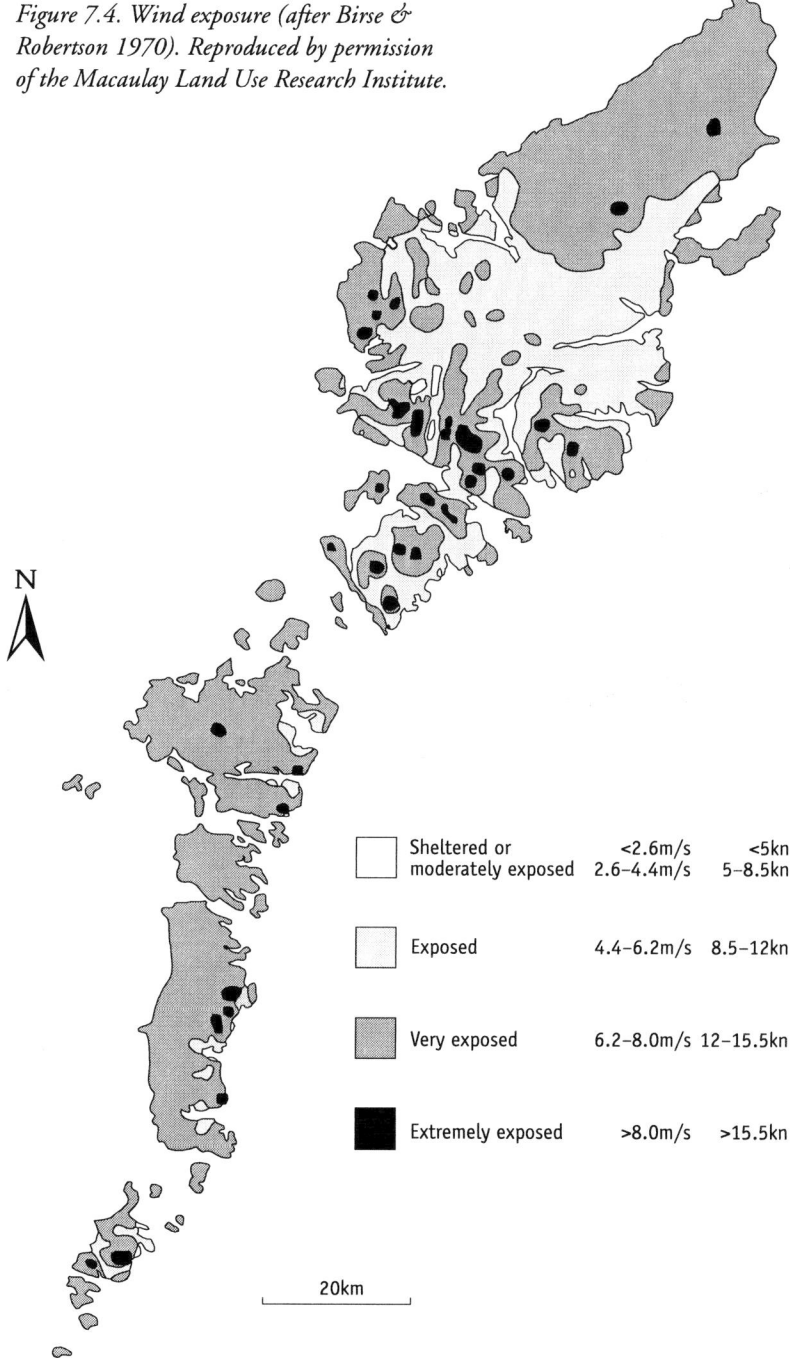

*Figure 7.4. Wind exposure (after Birse &
Robertson 1970). Reproduced by permission
of the Macaulay Land Use Research Institute.*

N

Sheltered or	<2.6m/s	<5kn
moderately exposed	2.6–4.4m/s	5–8.5kn
Exposed	4.4–6.2m/s	8.5–12kn
Very exposed	6.2–8.0m/s	12–15.5kn
Extremely exposed	>8.0m/s	>15.5kn

20km

Gaelic is rich in terms for different winds and forecasts for these, as catalogued by Raghnall Mac Ille Dhuibh[27]. A cold, bitter wind which first appears in January is the *gobag*, or biter, followed at the end of winter by the *faoilteach*, wolftime, which often brings a long period of snow. The *faoilteach* is followed by the *feadag*, or whistler (the same name is given to the golden plover), between mid-February and early March. The *gobag* or *sguabag* (sweeper) lasts between three and nine days, and is mainly associated with the periods 12–18 and 21–24 February, and 15–17 and 29–31 March

Martin Martin[32] claimed that in Harris the best time for fishing for salmon and trout was when a warm wind blew from the south-west. Fishermen, from the angler to the deep sea professional, have their own rules and requirements for the weather. One year, a colleague of local provenance was less than happy with the dry state of the rivers when salmon spawning was due to take place. He did a rain dance, which seemed to work, to the extent that by the time it had rained for forty days and forty nights, we begged him to perform the un-rain dance. Needless to say, he did not know that one, and the deluge continued. Martin also tells of a 'Water Cross' opposite St Mary's Church in North Uist which was erected by the "ancient inhabitants" to produce rain, and placed flat to bring the rain to a halt.

A possible whirlwind hit the North Uist island of Baleshare on 25 June 1958. It was seen approaching from the west as a dark spiral, "at a terrific rate and roar". As it crossed Baleshare, it "carried everything in its wake, ripped the tall grass and bushes from the roots ... and narrowly missed Donald Macdonald who was tending his ferry boat ... Following this a cloudburst of the greatest severity took place"[44].

SUNSHINE

Average daily sunshine is 3.40 hours at Stornoway and 3.89 hours at Benbecula. The slightly lower figure for Stornoway may be caused by a greater tendency for cloud formation over the land to the west and south-west (the sources of the prevailing winds), while exposed western coasts tend to have good sunshine records at sea level[20].

That the sunniest (and driest) months are May and June (Table 7.1) is related to the tendency for anticyclones to develop in The Atlantic during these months bringing fine, calm weather to the islands, sometimes for long periods[29].

The relatively high latitude of the islands means that days in midsummer are long, with very long twilights. Visitors delighting in the opportunity to peruse the columns of the *Daily Telegraph* outdoors at midnight in June, however, might spare a thought for the islanders who, in the dark, dismal days of December (average daily sunshine 48 minutes), require the assistance of artificial light to read their *Stornoway Gazette* at midday!

The high latitude also means that the sun's rays strike the islands at a relatively low angle, lengthening shadows and increasing the filtering effect of atmospheric haze[19]. Steep-sided, north-facing corries such as Coire Roineabhail in South Harris might be in perpetual shadow in winter, and receive very little sunshine even in summer.

Though measurements at Oban suggest that the north-west of Scotland receives almost as much energy from solar radiation as the south-east of England[20], the sun's heating effect in the Western Isles is severely attenuated by high winds and high humidity.

TEMPERATURE

The climates of Moscow (55°45'N) and Labrador (mainly south of the 55th parallel) are rather more extreme than that of the Western Isles, even though we lie a little further north than these areas, at a latitude of between 56°48'N and 58°31'N. That we have a comparatively mild climate for our latitude is almost entirely due to the fact that our islands lie in the path of the North Atlantic Drift, a warm ocean current which is an extension of the Gulf Stream. Every school pupil is taught that because of the Gulf Stream (and it should be the North Atlantic Drift) the waters of the Hebrides are warmer than those of the east coast of Britain. It must be said, however, that when bathers armed with this comforting knowledge venture into the Hebridean briny, protected (usually!) by only a brief costume, "warm" is rarely, if ever, among the range of adjectives which first spring to the lips. Only once have I known the sea to be warm enough to stay in for more than half an hour: Coll, in Broad Bay, is so gently sloping that the water remains shallow over large areas for long periods, very occasionally allowing the sea to heat up to the extent that even I could remain immersed for up to two hours without contracting hypothermia. When I was a teenager, I once saw swimmers (with no protection other than swimming trunks) entering Broad Bay in deepest winter, with ice in the saline pools on the shore; as an aspiring scientist, I concluded from the evidence before me that these people were quite mad. As the photograph of Loch Roag (Fig. 7.5) demonstrates, the sea occasionally freezes over, though this often reflects the presence of fresh water on the surface. In February 1958, swans were said to have been frozen alive in the ice at Carinish in North Uist[43].

Average surface sea temperatures off the Western Isles are 7–7.5°C in January and 13–14°C in August, keeping the air around the islands cool in summer and mild in winter. Summer sea temperatures around the Western Isles

Stewart Angus, February 1978

Figure 7.5. Frozen Loch Roag

are only about half a degree warmer than those on the North Sea coast of Scotland, but our waters are some 2.5°C warmer than those of the Moray Firth in winter[26].

The truly maritime nature of the climate of the Western Isles is reflected in the way that air temperatures tend to follow sea temperatures rather than the sunshine record. Thus the warmest months at Stornoway are July and August rather than the sunnier months of May and June, because the sea warms up (and cools down) very slowly.

The annual range in average monthly temperature at Stornoway is only 8.8°C, one of the lowest ranges in Britain – a true reflection of the ameliorating influence of the sea. The record maximum and minimum temperatures there are 25.6°C in 1911 and -12.2°C in 1960. A temperature of 27.2°C has been recorded in Benbecula[31]. The founder of the *Stornoway Gazette*, William Grant, complained in his column in the *Highland News* of 26 August 1899 that the heat in Stornoway over the previous two months had become "very oppressive", which cynics might suggest was the last time anyone was able to make such a remark.

On average there are 47 days of air frost a year at Stornoway, 33 in Benbecula, and an estimated 20 in Barra. The latest and earliest dates for frost are about 1st April and 1st December respectively in the Uists: nowhere else in Britain has a significantly longer frost-free period except the Scilly Isles and Channel Isles[19].

The 'lapse rate' or fall in temperature with increasing altitude varies according to season, from 1°C for every 167m ascent during October to December, through 1°C per 141m from July to September, to 1°C per 132m from January to June[47]. The figure is usually taken as 1°C per 167m, and on this basis the top of the Clisham (799m) would, on average, be 4.8°C colder than Stornoway. The higher winds at altitude and the resulting 'wind chill' would contribute to a much greater perceived temperature difference which should always be considered by hill walkers.

Until recently, it was generally agreed that for most plants growth was significant only above temperatures of 5.6°C. Potential growth at different localities was compared by measuring the number of days on which the temperature exceeded the growth threshold annually. For the record, Stornoway is above 5.6°C for 225 days a year[24]. This is not a particularly good measure of climate as it takes no account of the extent by which the temperature exceeds the growth threshold. In fact temperatures in the Western Isles are likely to be only slightly above 5.6°C for many of these 225 days.

A more useful yardstick of growing season is that of 'accumulated temperature', which takes account of the number of days on which, and degrees by which, 5.6°C is exceeded, expressing this measure as 'day°C'. In these terms, the warmest places in the Western Isles are in southern South Uist and Barra (Fig.7.6). Most of the land is 'fairly warm' and becomes cooler with increasing altitude, as would be expected. Only the higher summits are 'very cold', and there are no 'extremely cold' areas with annual accumulated temperature lower than 300 day°C[6].

It is now known, however, that many plants grow at temperatures as low as 0°C, albeit slowly, and it is thought that higher temperatures are more important in the earlier half of the year. New figures for accumulated temperature have been calculated using 0°C as the growth threshold and covering the months January to June. The values of this 'lower quartile of accumulated temperature' are 1179 day°C at Stornoway, 1265 at Balivanich, and 1311 in Tiree (cf. Lerwick 954, Kirkwall 1073, Wick 1066, Inverness 1246, Dumfries 1249, Cambridge 1413 and Penzance 1658)[5]. As Fig. 7.6 was originally based on the annual accumulated temperature for only a few stations, the distribution of the categories of 'lower quartile accumulated temperature' will be very similar – only the figures will be different.

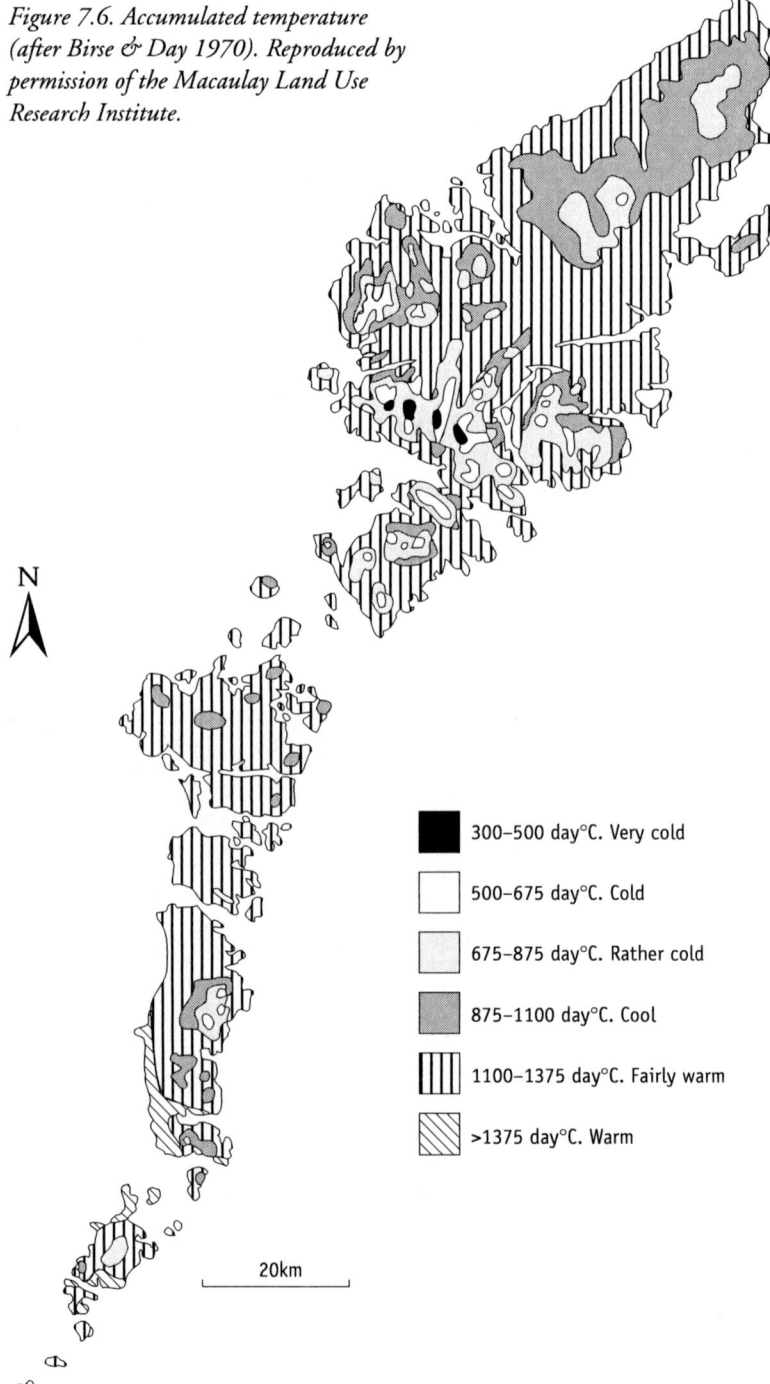

Figure 7.6. Accumulated temperature (after Birse & Day 1970). Reproduced by permission of the Macaulay Land Use Research Institute.

PRECIPITATION

The average annual precipitation (i.e. rain, sleet, snow and hail recorded in liquid form) at Stornoway is 1094mm (43.1") and is 1220mm (48") at Benbecula Airport[31]. Some 1200-1400mm fall annually over most of the Uists and northern Lewis (Fig.7.7). Dew, hoar frost and fog also contribute moisture but cannot be easily measured in rain gauges, and for this reason are known as 'occult precipitation'[19].

Where the prevailing winds force cloud upwards over mountains and hills, the lower temperatures at these levels cause condensation, resulting in a fall of 'orographic' rain. Thus most of southern Lewis, Harris, and the Beinn Mhor range of South Uist receive 1400-1600mm or more annually, while some of the western summits of North Harris receive over 2400mm. This is more than double the total for Stornoway, and 800mm more than nearby Hushinish receives at sea level.

While moist air is cooled and condensed by ascent on the windward side of mountains, it is warmed and dried by its descent on the other side. Thus the north and east of Lewis, being on the leeward side of the high ground in relation to the prevailing winds, are said to be in the 'rain shadow' of the hills and receive less rain than falls at comparable altitudes on the west coast of southern Lewis and North Harris. Stornoway and Point are in the rain shadow of the south and west of the island, and parts of the Broad Bay area receive less than 1100mm of rain annually, with 1011mm recorded at Back School[31].

Rainfall is fairly evenly spread over the year, the number of days with measurable rain per month ranging from 17 to 26, and totalling 263 over the year (Table 7.1). Of these 263 days, 200 are 'wet', with more than 1mm of rain recorded[41]. December is the wettest month and May is the driest: 30% of the rain falls in winter (Dec-Feb), 19% in the spring (Mar-May), 21% in the summer (Jun-Aug) and 30% in the autumn (Sep-Nov).

Blanket bogs receive all their nutrients from rain water. Studies elsewhere in Britain[1] suggest that the amounts of nutrients in rain usually rise with increasing rainfall. In the Western Isles the proportions of nutrients in rain are very close to those in sea water[23]. Acid rain is not known to have reached critical levels in the Western Isles, and the pH of Hebridean rain rarely even approaches 4.5, when aluminium in the soil would become chemically available and could prove toxic to young fish.

Average relative humidity is high throughout the year, ranging from 80 to 88% saturation, with an annual mean of 85% (Table 7.1).

Though snow is observed falling on 47 days a year at Stornoway, it is lying (at 0900hrs) on only 11 of these days (5 in the Uists). The maritime climate of the islands means that snow seldom lies for more than a few hours, though

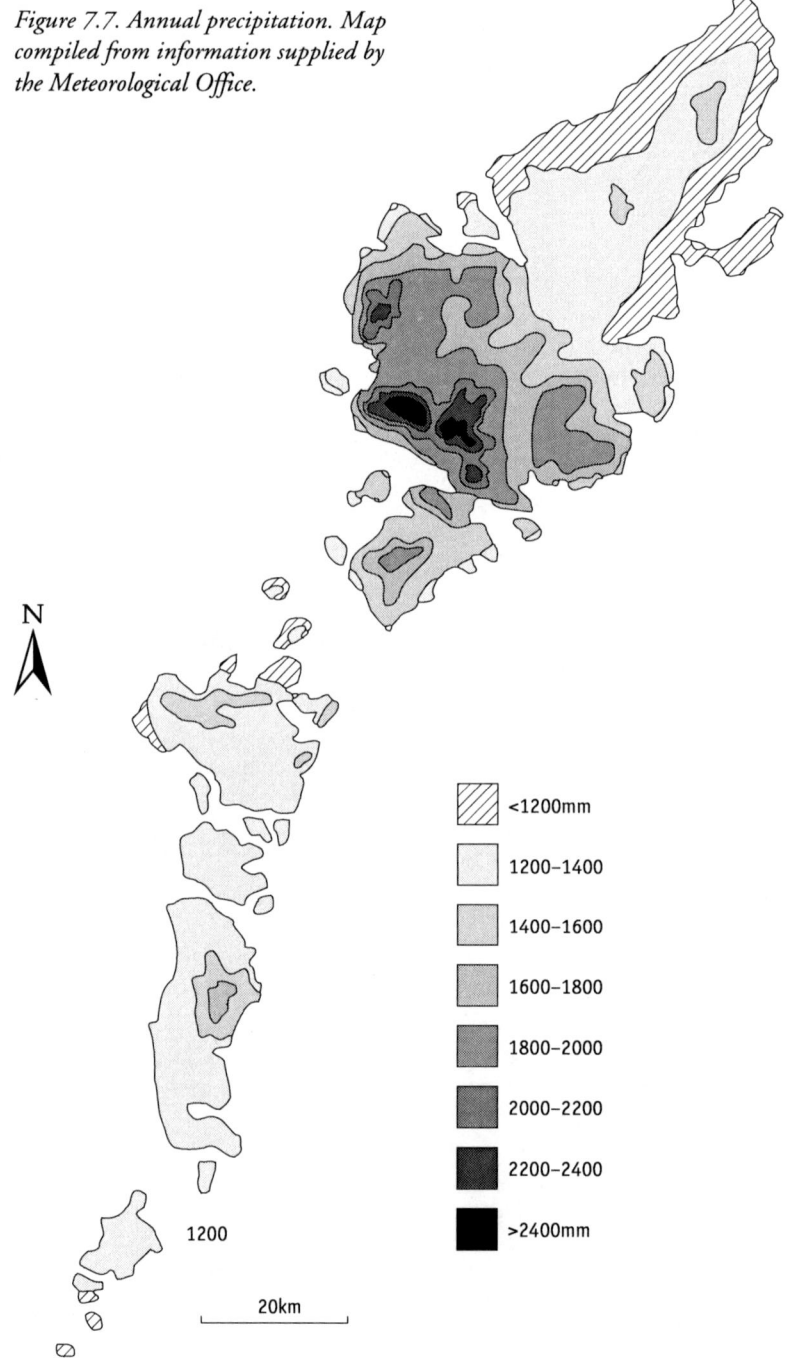

Figure 7.7. Annual precipitation. Map compiled from information supplied by the Meteorological Office.

on the higher hills, where air and ground temperatures are lower, snow cover is more persistent, especially in north-facing corries which receive no winter sun and relatively little direct sunshine even in summer. Snow may also lie later in block screes, which also tend to be located below north-facing slopes, as they were mostly formed by freeze-thaw action. There are none of the really late snowbeds associated with the higher mainland ranges.

Remarkable meteorological phenomena involving snow were observed at Stornoway Airport on 8th February 1954, when a plane seemed to "create its own fog". During the early afternoon, visibility was excellent, conditions were almost calm, and snow, lying to a depth of some 12-15cm, was 'steaming' in the sunshine. The temperature was -1.7°C. The aircraft, a British European Airways Pionair, began to taxi for take-off at 1457hrs. As the engines reached full power, snow was thrown into the air, and fog began to form where the snow settled. Within quarter of an hour, the entire airport area was enveloped in fog to a depth of 4-11m, and the temperature had plummeted to -10°C, the lowest recorded at Stornoway since 1945! It is possible that these conditions were caused by very rapid mixing of cold surface air with warmer, damper air above, with the air being cooled by the snow particles, though the temperature of the snow on the ground was only -7.8°C[15].

Soft hail sometimes falls from cumulo-nimbus clouds on winter cold fronts, and is often associated with snow. An exceptionally severe hailstorm, with hailstones up to 25mm in diameter, was experienced in the Cnip, Timsgarry and Miavaig areas of Uig on 29 October 1986, damaging the windows of houses and the bodywork of vehicles.

Maritime climates in northern latitudes are characterised by the low incidence of thunderstorms, with a greater tendency towards cloud-to-earth rather than cloud-to-cloud discharges[19]. There are on average three days with thunder each year, two of which are usually associated with winter cold fronts[31].

On the flat, treeless landscape of northern Lewis, anything projecting more than a few feet above the surface of the land has a chance of being struck by lightning during a thunderstorm. In Uig, a woman was killed by lightning in 1938[45], and in April 1976, four young people who were walking home in the early morning on the west side of Lewis were thrown to the ground by lightning, one of whom was tragically killed[46]. One of the worst thunderstorms in the history of the Outer Hebrides is said to have been that of August 1897: according to the *Highland News* of the time, the lightning split rocks in several places, and "the rain fell in torrents, accompanied by snow and hail" (in August!). George Macaulay of Hacklete was thrown three times in succession, and others felt as though they had been struck on the back by a spade when lightning hit nearby peat banks and tore the ground.

Lightning striking dry vegetation may cause a fire, and there is evidence that such fires occurred before the first humans arrived in the Hebrides.

In *The Gaelic Vikings*, James Shaw Grant[17] tells the story of the Tolstachaolais Poltergeist. One Sunday morning, an old lady who lived in a wooden house atop a rocky spur in Tolstachaolais was having a cup of tea with her grandchildren. Without any warning, caorans (small peats) began to jump from the cast iron stove, one of them landing in her tea, then crockery began to move, some of it smashing into walls, other pieces shattering where they stood. Grant pointed out that this incident had taken place during a time of great electrical activity over Europe and that the iron stove, surrounded by the insulation of the wooden house, could have built up a massive charge which attracted the peats and the crockery.

Electrical disturbances have been invoked to explain some other strange phenomena in the islands. Around the time of the Tolstachaolais Poltergeist, a crofter was struck by lightning. Though he was unhurt, his trousers are said to have been ripped off. The man must have been extraordinarily unlucky, for when he looked in a trunk for some of his other clothes, he found that they had crumbled[17]. Personally, I blame moths!

To this catalogue of strange phenomena, we must add one more incident: a story which seemed so strange at the time that even now, nearly forty years later, and despite the viable explanations offered at the time, people still talk of it with unease, though perhaps suspecting a human rather than an extraterrestrial explanation.

The year is 1958, and the scene Loch nan Learga, between Morsgail and Kinlochresort in Uig. The loch is too small to be named on the old one-inch Ordnance Survey maps, but is clearly outlined at NB128193. You will look in vain for this loch on the later 1:50,000 maps, or even the larger scale 1:25,000 sheet because, on 20th November 1958, Loch nan Learga disappeared.

Later in the month, a "plane load" of reporters visited the site of the incident, which was immediately headlined as the "Morsgail meteorite", attracting sufficient press attention to merit a CBS camera team and a mention in the *Bulawayo Chronicle*. The nationals of 1st December, notably the *Daily Express*, carried aerial pictures of the area, clearly showing the drained loch, with the bedrock below the loch scoured clear of peat, and the peat below gullied to a depth of 1.5m. It was clear that Loch nan Learga had emptied into the lower-lying Loch Mór Sheilabrie; the latter loch was still turbid with peat, and had an alluvial fan of peat denoting the location of the inflow of water. Additionally, the outflow from Loch Mór Sheilabrie showed signs of a flow surge.

Though the experts who had been flown to the site offered a logical explanation, this was much less exciting than a meteorite, and local people are said to have found problems with the explanation of heavy rainfall having led to a 'bog burst', on the grounds that the rainfall had been no heavier than is usual for winter, and this was the only site involved in an area studded with similar lochs.

D.R. Bowes, one of the scientists involved, wrote an account[10] of the incident in the *Scottish Geographical Magazine* which explained the reasons for the bog burst. The two lochs were on peat overlying Lewisian gneiss, like most of the smaller lochs in peaty areas in Lewis, but between the two was a ridge of amphibolite. The key to the incident was not the heavy rainfall, but the unusually dry early summer, which had created desiccation cracks in the peat. As any peat cutter knows, peat which has dried out never recovers its natural state, no matter how much it is soaked, and the heavy rain of the winter soaked through the cracks, creating a layer of sludge above the bedrock. The peat between the ridge and Loch Mór Sheilabrie was the first to slide downhill using the lubricating effect of the sludge; without the security of the weight of this peat, the shallower peat above the ridge which had impounded Loch nan Learga gave way suddenly under the weight of water, easily breaching the sub-peat amphibolite ridge, assisted by the underlying sludge (Fig. 7.8). The force of the water surge was such that blocks of peat were deposited above the original level of the loch.

Large bog-bursts occurred on steep north-facing slopes on Barra and Vatersay in 1992[14].

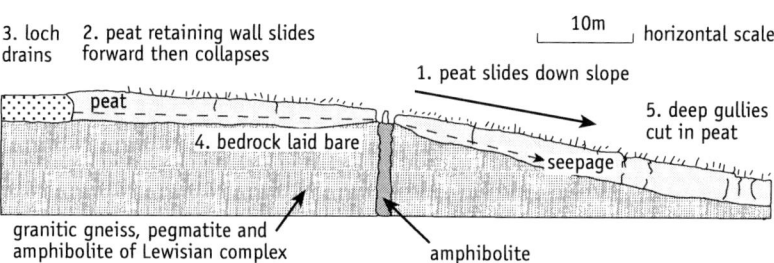

Figure 7.8. Cross-section of soils at Loch nan Learga (after Bowes 1960)

Though it would be easy to give a map of precipitation and leave it at that, it is really how rainfall (and snow and hail) behave after reaching the ground that is important to plants and animals – and to land users. Wind speed and direction, slope, vegetation type and soil type and temperature at the time will all combine to determine how plants will be affected by precipitation.

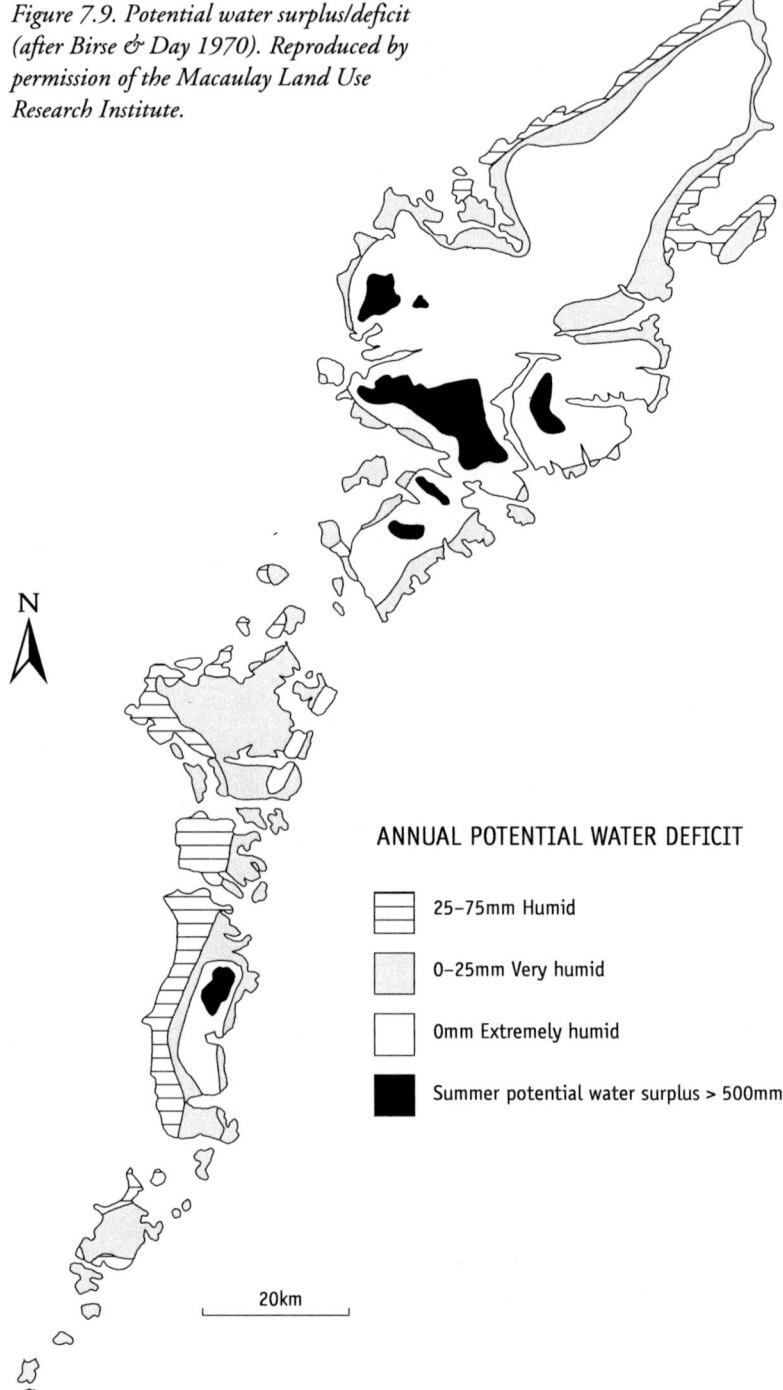

Figure 7.9. Potential water surplus/deficit (after Birse & Day 1970). Reproduced by permission of the Macaulay Land Use Research Institute.

N

ANNUAL POTENTIAL WATER DEFICIT

25–75mm Humid

0–25mm Very humid

0mm Extremely humid

Summer potential water surplus > 500mm

20km

The ecological significance of precipitation figures in terms of soil moisture may be judged by relating them to the amount of moisture returning to the atmosphere by direct evaporation from soil and plant surfaces, or taken up by plant roots and returned to the atmosphere by evaporation from pores in the leaves, a process known as transpiration. If, over a period, rainfall exceeds the potential water loss from evaporation and transpiration, there is said to be a potential water surplus (PWS), as indeed there is over most of the year in the Western Isles. For some periods of the year, potential transpiration and evaporation losses may exceed rainfall by an amount known as the potential water deficit (PWD). Only the higher ground of Lewis and Harris and the Beinn Mhor range of South Uist have a summer PWS exceeding 500mm. In terms of PWD, the Uists are less humid than Lewis and Harris, and have the largest 'humid' area (Fig.7.9). This is related to the fact that rainfall increases with altitude, while evapo-transpiration decreases[8].

While it should be pointed out that PWD may not be as important to wild plants as it appears to be to grain crops[7] and that *actual* water deficit is difficult to measure (so that *values* given may not be accurate), PWD is a useful indicator of the *pattern* of distribution of climatic limits on vegetation[19].

Stewart Angus, October 1987

Figure 7.10. The water surplus: large areas of surface water on peatland near Loch Langavat, Lewis, near the border with Harris

OCEANICITY

Continental climates become increasingly maritime or 'oceanic' westwards, an effect superimposed on the expected influence of increasing latitude. With increasing oceanicity, precipitation increases and is more uniformly distributed over the year, the number of 'wet' days increases, and humidity and wind speeds rise, while annual temperature range decreases[38,34].

All the islands of the Western Isles are classed as 'hyperoceanic', this being the most oceanic sector of Eurasia[6]. St Kilda, with a summit of 426m, is high enough to develop its own localised weather conditions, with increased rainfall, humidity, cloud cover, and wind exposure[11].

The extreme oceanicity of the climate of the Western Isles has a profound effect on vegetation, so that montane plants grow at much lower altitudes than in more continental areas such as the Cairngorms. In the Western Isles and on the north-west mainland of Scotland it is not uncommon to find 'mountain' plants growing at or near sea level, a phenomenon known as 'altitudinal descent'.

Box 7.1 – Plants showing altitudinal descent in the Western Isles
(Angus 1991)[2]

Purple saxifrage	*Saxifraga oppositifolia*
Alpine bistort	*Polygonum viviparum*
Mountain sorrel	*Oxyria digyna*
Roseroot	*Sedum (=Rhodiola) rosea*
Moss campion	*Silene acaulis*
Hoary whitlowgrass	*Draba incana*
Bearberry	*Arctostaphylos uva-ursi*

The high oceanicity results in later flowering of many plants in relation to their counterparts elsewhere.

CLIMATIC HISTORY

Climatic trends since the departure of the last glacial ice from Britain are the subject of considerable discussion among climatologists and ecologists.

Some clues to the nature of past climate can be gleaned from the study of pollen preserved in loch sediments and peat. Cores are extracted from the sediment, and samples are taken from a number of levels, some of which are Radiocarbon dated to provide reference points in time. Pollen is then identified and counted at each sample level, and a pollen 'profile' of the sediment is drawn. Changes in the occurrence or abundance of different species with time give clues to climatic change over the period which the profile represents: some plants suggest wet conditions while others might prefer warm temperatures.

Scientists have used the evidence of widely available pollen profiles in conjunction with information gained from seabed cores, ice cores and tree rings to reconstruct the pattern of Postglacial climate.

The *Pre-Boreal Period* dates from the end of the last glacial phase, around 8300 BC, and lasted until about 7000 BC. It was a time of rising temperatures and over this period tundra vegetation types were increasingly replaced by plants more characteristic of temperate conditions[25].

The temperature rise continued into the *Boreal Period* (c.7000 to 5000 BC) and into the *Atlantic Period* (c.5000 to 3000 BC). The Atlantic Period is sometimes referred to as the 'Postglacial climatic optimum' and summer temperatures may have been 2-3°C warmer than they are today, though rainfall may have been as much as 15% higher, at least in England[25,21].

The *Sub-Boreal Period* (c.3000 to 1000 BC) was marked by recurrent fluctuations in climate, and is thought to have brought increased wetness, more acid soil conditions, and possibly stronger winds to the north of Scotland between 2600 and 1600 BC[25].

This was succeeded by the *Iron Age Cold Period* or *Sub-Atlantic Period*, which lasted until around 300 BC (later in many areas). (Iron Age culture did not reach Scotland till around 250 BC). The temperatures of the cold period were about 1°C lower than at present, and rainfall was probably higher[21].

The *Secondary Optimum* spanned the years 800–1300 AD[25] and summer temperatures were about 1°C warmer than they are now. This amelioration in climate is said to have allowed the great Viking expansion[21], which had reached Lewis by the mid-ninth century[28].

The *Little Ice Age* lasted from about 1430 till 1850. The Arctic pack ice extended southwards, and temperatures were some 1–3°C lower than they are today[21]. Trouble with drifting sand at Udal in North Uist seems to have begun around 1400 AD, with calmer conditions returning in about 1500. Stormy conditions had returned by 1542, and the gale of 1697 was so severe that the site was abandoned forever, after almost 4000 years of continuous occupation[3].

The occurrence of pumice at several archaeological sites suggests that this porous volcanic material floated to the shore of the Western Isles (though it could conceivably have been imported from elsewhere by Man) where it was used for shaping and polishing bone and wood[4]. Though the pumice has been linked to the eruption of Hekla in Iceland in 1510 AD this origin has now been discounted, but the 1510 eruption was responsible for the atmospheric dispersal of tiny needle-shaped particles of volcanic material known as *tephra*, which can be identified in many deposits in the Western Isles. These materials are important as indicators of major eruptions which could have had a significant effect on the climate of north-west Scotland, possibly restricting light and even leading to the temporary acidification of fresh water lochs and the death of their fish populations, but there is as yet no definite evidence of such events in the Western Isles[12,18].

Reference has already been made to the years of the kelp boom, and the effects on population and cropping (see Chapter 5). The higher population was sustained largely by the potato. From the late 1840s, several successive years of warm, damp weather with light winds coinciding with the potato growing season provided ideal conditions for the spread of the potato blight fungus *Phytophthora infestans*, causing a devastating famine. Though the consequences were not so serious as in Ireland, they were nevertheless severe.

More recent variations in climate were suggested by the finding of ice skates in the wall of a house at 83 Keith Street in Stornoway during renovations in late 1994. The house had been built about 100 years before, and initially it was speculated that someone had brought the skates home from Canada during the Depression. However, the *Highland News* of 23 January 1897 reported that a curling club had been set up in Gress, using "Mr Liddle's mill dam" as their rink. Two years earlier, the same paper revealed that "as nearly all the lochs [in Lewis] are frozen with a thick coating of ice, skating has been much indulged in." It would be rare today for a loch of any size to be frozen so thickly that it could be walked upon by an adult.

In Harvie-Brown's diaries for 1886, he refers to Captain Macdonald of Rodel's recollection of the time when the people were able to make salt in shallow rock pools by evaporation, suggesting prolonged periods of warm, dry weather, though possibly just the early summer period of high pressure mentioned above.

Some authors dispute this sequence of climatic fluctuations, particularly the Atlantic-Sub-Boreal transition, which is widely held to have been responsible for the onset of blanket bog growth and the decline of tree cover. The eminent biologist Dr Burkhard Frenzel[13] maintains that average Postglacial summer temperatures in Europe oscillated by only about ±0.7°C, while the geographer R.J.Price[39] has warned of the dangers of applying even English climatic history to Scotland. Professor Gordon Manley[31], writing of the Outer Hebrides, has said that "There is little to suggest that the minor climatic fluctuations in historic time, and even those of the postglacial, have ever been of sufficient amplitude to make a serious difference to the climate from the human standpoint". Work in the Cairngorms, however, has shown that trees once grew at altitudes where tree growth would be impossible today, which is most easily explained by a deterioration in climate[36,42]. A change to wetter, windier conditions about 4000 years ago is also indicated by analyses of sediments from lochs on the north-west mainland of Scotland[37].

Climate has clearly had a major impact on soils and agriculture; this is dealt with in the next Chapter.

8

AN INTRODUCTION TO SOILS

===➤·0·◄===

Lewis is essentially a large peat bog which has on its fringes, the seaboards of the Minch and of the Atlantic, stretches of eroded or 'skinned' peat. Almost all the crofts in Lewis lie on the coast and so all the cultivation is on this shallow peat or mineral soil. It is almost all acid soil. The patches of machair are few and small. (Grant 1979)[12].

Soils are not easily described: their classification relies on an understanding of the importance of a range of 'horizons' (layers), their sequence and colouring, as well as an ability to relate the soil profile and its development to topography, climate, geology and landform. Though there are fewer technical expressions than in geology, full definitions would require more space than is available here.

The best introduction to the soils of the islands is undoubtedly that by Gordon Hudson in the *Flora of the Outer Hebrides*[29], but for the full picture, you must refer not only to the superb Handbook on the Outer Hebrides by the Soil Survey of Scotland[16] and its accompanying 1:250,000 map, but also to the explanatory Handbook for the Soil Survey series[33]. There are also outline (unshaded) maps at a scale of 1:50,000 published by the Macaulay Institute for Soil Research (now the Macaulay Land Use Research Institute – MLURI).

Soil is usually a complex material consisting of weathered minerals derived from local or transported rock debris, humus or organic matter (decayed or partially decayed plant material), air and water. The type of soil which develops in a particular location depends on the nature of the parent material, slope, climate, time, and on the influence of vegetation and human activity[10]. Soil may also be affected by the organisms which live within it.

Where the parent material is mineral-rich, then the soil also tends to be mineral-rich, but certain provisos apply: prolonged, heavy rainfall can leach minerals out of the soil, and some soils may bear no relation to the underlying rocks, having been transported there from their original 'parent' area by glaciers as till (see Chapter 3) or by water as alluvium, sands or gravel.

The Lewisian rocks which make up so much of the island landscapes are not only generally resistant to weathering, but yield little other than quartz and feldspar in their glacial debris, and therefore form acid, nutrient-deficient soils. Only the sedimentary rocks of the Stornoway Formation (and the tills derived from these) contain more promising materials. The tills of Ness, consisting of material carried from the bed of the Minch (see Chapter 3) are more fertile.

A typical altitudinal sequence of soils in the Outer Hebrides (outside machair areas), beginning at sea level would be: humus-iron podzols, peaty podzols and peaty gleys, subalpine podzols and subalpine gleys[15]. In parts of Harris, Benbecula, and North and South Uist, the soil pattern is very complex, and several of the soils described below occur in very close proximity to each other, often in a recurring pattern across large areas (Fig. 8.1).

IMMATURE SOILS

In areas where the climate is severe, as on mountain tops, frost-shattering and transport of debris downwards by gravity have produced screes and other deposits of rock fragments which support little vegetation, and can scarcely be described as soils, having little in the way of horizons or overlying relatively unaltered parent material. Where vegetation has become established on this poor substrate, the resulting layer of humus, up to 10cm thick, lying directly on the rock or rock fragments, forms a *lithosol* (Fig. 8.2). If the humus layer exceeds 10cm in thickness but still rests directly on the rock fragments, the soil is a *ranker*. Rankers in the Western Isles are often peaty[16].

Stewart Angus, March 1991

Figure 8.2. Lithosol on the plateau of Roineabhal, South Harris

*Figure 8.1. Soil map
(after Hudson 1991₁₄)*

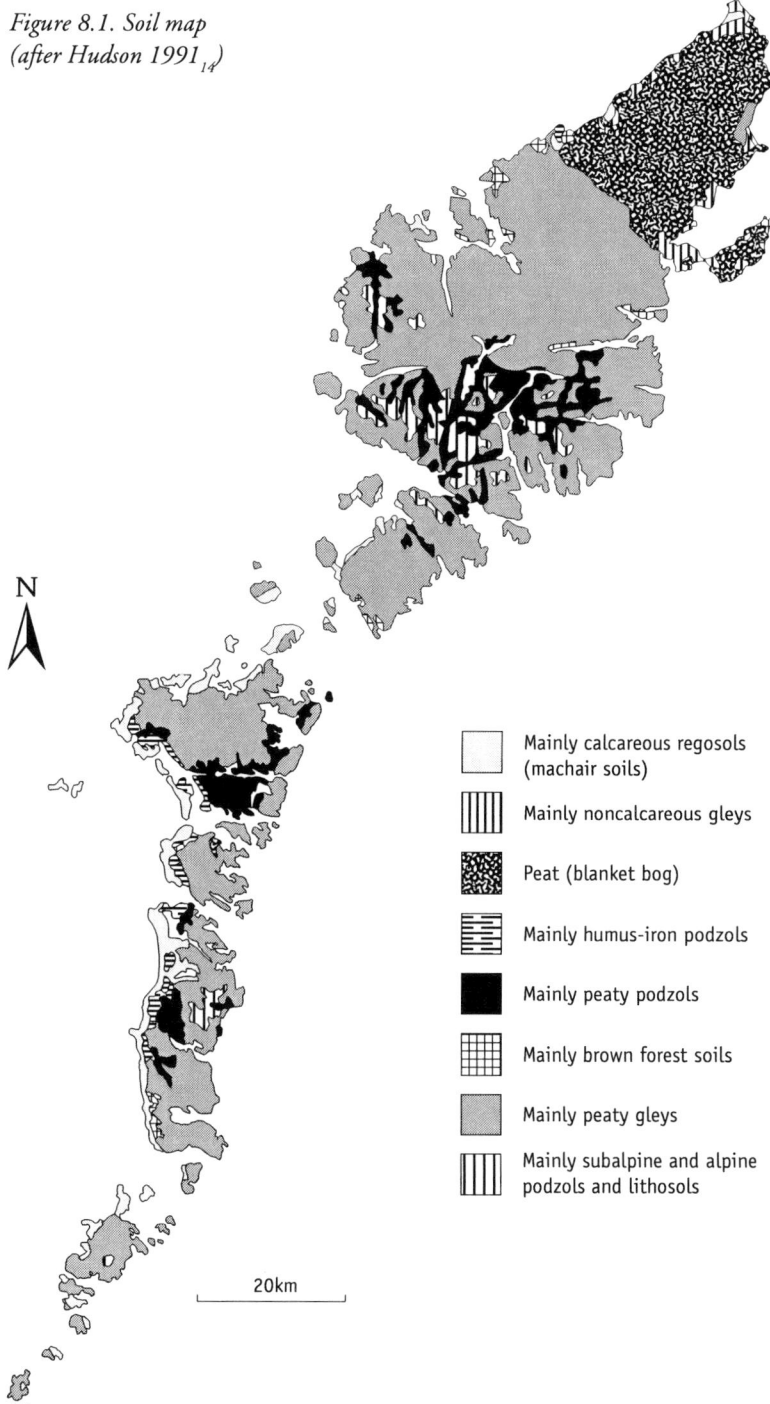

N

Mainly calcareous regosols
(machair soils)

Mainly noncalcareous gleys

Peat (blanket bog)

Mainly humus-iron podzols

Mainly peaty podzols

Mainly brown forest soils

Mainly peaty gleys

Mainly subalpine and alpine
podzols and lithosols

20km

The huge quantities of fluvioglacial sand which, along with shell fragments, have been washed and blown ashore since the end of the last glacial period, have accumulated on land as the *regosols* of the machair (Figs 6.7, 6.10). Only on the more stable, long-established machair areas is there an appreciable humus content, indicated by the darker colour of the soil. Former soil surfaces are often readily apparent in machair soil profiles exposed by erosion as darker layers which are richer in humus (Fig. 8.3). Archaeologists find the old soil layers or *palaeosols* vital in their interpretation of machair history[9]. The archaeologists and other scientists associated with the Sheffield Environmental and Archaeological Research Campaign in the Hebrides (SEARCH) have described soil layering in machair grassland as *machair stratification*, made up of very fine, alternating layers of organic material and blown sand[9] which confirms earlier work by Professor William Ritchie (see also Chapter 4).

Stewart Angus, May 1995

Figure 8.3. Organic layer in eroding machair margin, Pabbay (Sound of Harris). Note slumping on right, illustrating strength of root system.

At Tràigh Teinish in Uig, and on the island of Pabbay in the Sound of Harris, the surface of the machair soils may become 'crunchy' in places because of a crust of calcareous material on the surface. The process at work here is virtually the same as that responsible for the formation of the stalactites in the Seal Cave at Gress (see Chapter 2). The carbonate in the shell sand (calcium carbonate) dissolves in acid water in the absence of oxygen underground, becoming soluble bicarbonate. On returning to the surface, the bicarbonate once again becomes insoluble carbonate, forming a solid deposit on the surface of the soil. It was possibly this deposit which was described by Pochin Mould[21] on Barra, when she said that "The shell sand of Pabbay is so rich in lime that with the frequent rain it has dissolved and been redeposited to form beds of impure limestone".

Alluvium, material deposited by water, is of three types: saltmarsh, freshwater, and peaty. Rivers usually carry little silt in the Western Isles because of the unyielding rocks, while peaty alluvium tends to be restricted to the margins of lochs. Saltmarshes are described in Chapter 4.

LEACHED SOILS

Thousands of years of high rainfall, low evaporation, and the resulting water surpluses, have meant that the surface layers of many soils of the Hebrides have been 'leached' of their few nutrients by downward-percolating water, creating exactly the right conditions for the build-up of organic material (peat). Only in areas where the parent material has been particularly rich, as in the sea-borne till of Ness, are there unleached *brown calcareous* soils[14].

Where iron and aluminium have been leached downwards, they accumulate below, forming a *humus-iron podzol*, a soil that exists today only where drainage is better than average and man has not disturbed the soil by cultivation[16]. Humus-iron podzols form most easily on permeable parent materials[15] such as fluvioglacial gravels or tills derived from the rocks of the Stornoway Formation.

In peaty areas, there may be high concentrations of ferrous salts dissolved in surface water which may be precipitated (deposited) as a conspicuous hard 'iron pan' of ferric compounds[30]. This can often be seen as a dark, undulating line in soil profiles exposed in road cuttings and quarries. The pan impedes the downward percolation of water, so that it accumulates in the leached layer, encouraging the formation of peat on the surface, and the soil is classified as a *peaty podzol*. In upland areas there are also *sub-alpine podzols* and *alpine podzols*, the latter where freeze-thaw processes (see Chapter 3) still operate[16].

Although leaching occurs in soils developed on richer parent materials the effects on the profile are less dramatic, and *brown earths* develop. These tend to be restricted to the better non-machair crofting townships such as Breasclete, Bragar, Leurbost, Rodel and Carloway.

WATERLOGGED SOILS

Partially or permanently waterlogged soils are often anaerobic (lacking oxygen). *Gleys* develop over compact, relatively impermeable tills[15]. The blue-grey appearance of the soil is associated with ferrous iron compounds which form in the absence of oxygen; these contrast with the reddish brown hues associated with the ferric compounds which form in aerated soils such as podzols. Peaty gleys are often found around the fringes of the deep peat areas of Lewis.

ORGANIC SOILS

Peat readily accumulates on podzols, gleys and rankers to form peaty versions of these soil types, but only when the organic layer is deeper than 50cm is the soil known as *peat*.

Peat is defined by the Macaulay Land Use Research Institute as "an organic soil which contains more than 60% organic matter and exceeds 50cm in thickness"[16]. The minimum depth specified is admittedly arbitrary, and some authors have placed it as low as 30cm[5].

As far as is known, blanket bog is of more recent origin than valley or basin-type mire, both of which may date back up to 10,000 years. The existence of tree roots which have yielded radiocarbon dates as recent as 3190 BP[34] and numerous archaeological remains beneath the peat confirm that some, at least, of our blanket peat is of comparatively recent origin. The only radiocarbon date so far obtained for the onset of blanket peat formation was obtained from an exposed site at Sheshader, Point, where peat formation began around 1700 BC (c. 3700 years ago)[6]. Dates for archaeological remains below peat range from around 4300 BP at Rosinish, Benbecula, to about 2800 BP at the Calanais Stones[6].

The partially decayed plant remains which constitute the peat provide a record of past vegetation, though with increasing age the remains are more completely humified and are thus more difficult to identify. This record is, however, more or less confined to the main peat-forming plants. Sphagnum mosses seem to be the major peat-forming plants in the Western Isles, though significant contributions are made by Deer-grass *Trichophorum cespitosum* and Hare's-tail Cottongrass *Eriophorum vaginatum*[8]. On St Kilda, there is unusual peat formed from plantains *Plantago* species, chiefly around An Cambir and at the tip of Ruaival. The St Kilda peats are even more unusual in that they support an earthworm fauna[13].

A more valuable record can be obtained from the pollen grains preserved in the peat, for pollen grains are extremely resistant to decay, and analysis of these can yield an insight into the nature of past vegetation over thousands of years. Pollen analysis is particularly valuable in investigations into the history of woodland.

True peat extends from sea level up to about 400m, reaching this maximum altitude just west of the summit of Gormol in Pairc (at approximately NB294067) but tends to be replaced by peaty soils on the low, isolated hills which rise above the Barvas Moor plateau[16]. Generally, though, blanket peat is associated with flat and gently sloping areas.

Peat depth is closely related to the underlying topography, with deeper peat filling depressions and shallower peat on slopes[11]. The average depth of peat on the Barvas Moor is 2.1m, with a maximum recorded depth of 5.1m[8]. Valley bogs may be as deep as 10m, but these are confined to small areas.

MAN-MADE SOILS

The only extensive area of *brown forest soil* which is found in the Western Isles occurs in the Stornoway Woods (with humus-iron podzols)[16], but it is possible that these soils were imported from the mainland by Sir James Matheson[4]. Local tradition has it that soil was also imported to Borve Lodge in Harris by Lord Leverhulme, and that the soil originated from Ireland.

Soils in many areas have been substantially altered by human activities, notably the conversion of cut-over peatland to pasture on the 'skinned land' of northern Lewis and where *feannagan* or lazybeds have been constructed (Fig.8.4). The description of *feannagan* by Fraser Darling[7] cannot be bettered. As Darling

Stewart Angus, November 1990

Figure 8.4. Aerial view of feannagan around the old village on North Rona

himself observed, there was a dramatic contrast between this form of agriculture and that practised on the machair.

> Considering the steepness and rocky nature of the slopes, and that such soil as can be scraped together on them is derived from peat and applied seaweed, it is amazing that any serious agriculture is followed at all. Nearly all the tillage is done on lazybeds or *feannagan*, platforms built up on the rock and in the hollows to provide as much drainage as possible for the soil. Potatoes and oats are grown on these with great care. Nothing can be more moving to the sensitive observer of Hebridean life than these lazy-beds of the Bays district of Harris. Some are no bigger than a dining table, and possibly the same height from the rock, carefully built up with turves and the seaweed carried there in creels by the women and girls. One of these tiny lazy-beds will yield a sheaf of oats or a bucket of potatoes, a harvest no man should despise.

Clearly the English name is misleading: Bill Murray, in his outstanding *The Islands of Western Scotland*[22] explains that the Gaelic is from *feann*, to scarify, while the English is from ley, meaning untilled ground, which does not sound all that convincing, given the amount of work involved in maintaining the ground as arable. In winter or evening light, the distribution of *feannagan* on ground which is seemingly unfit for cultivation is striking, and evokes a picture of a time when hunger drove local people to the limits of the land's ability to provide. It is possible, however that some, at least, of the ridges in the more unlikely areas, reflect not cultivation but attempts at pasture improvement, and Bill Lawson[17], the Harris genealogist and land economist, has cited the higher ridges at Nisisidh, near Scarasta, as a possible example of this, on the grounds that the highest ridges there are unlikely ever to have provided crops even in times of better climate.

There are conspicuous green ridges on low-lying ground just to the north of Loch Laxdale near Luskentyre NG108963, which are plainly visible from the road. They do not look quite right for *feannagan*, and I asked the Factor for Borve Lodge Estate in South Harris, Tony Scherr, if he could throw any light on the matter. He informed me that they had been built by the seventh Earl of Dunmore in the 1850s as part of an early fish ranching scheme. After stripped salmon eggs had been fertilised and hatched in a hatchery, the channels between the ridges were flooded and the very young fish were kept there as fry prior to release to the wild. There is a more extensive series of similar ridges along the banks of the Torray River in Barvas, at NB367487, NB369486 and NB374487 (Fig. 8.5). Ridges near a stream at Cleit a'Chragain, Pairc NB3304, may be similar structures.

The use of the alkaline shell sand of the beaches and machairs to neutralise or 'sweeten' the acid peatland soils of the interior dates back to at least 1808, when Macdonald reported in his *General View of the Agriculture of the*

David Maclennan, 1997

Figure 8.5. Man-made ridges, probably spawning channels, on the banks of the
Torray River, Carloway, Lewis

Outer Hebrides[19] that he was "gratified to see, in the remote district of Harris in
the Long Island, a sloop laden with shells and shell sand, which she had imported
from Barray [Barra]. … This proved to be the ninth [voyage] made by the same
vessel for this kind of manure within the four months preceding July 1808. The
cargo, about 40 tons, top-dressed three acres, and manured one acre completely
for oats." Professor Walker had certainly recommended the use of shell sand for
such fertilisation as far back as 1771, and his advice seems to have been heeded
by Lord Macdonald who, in 1799, instructed J.Blackadder to: "Examine the
shell sand which you will find breaking out on the beach in beds from one to
three feet thick, consider and point out where it can be carried inland to fields
of dry land, either by roads or canals. It is an excellent manure."[20]. The word
'manure' tends to be used nowadays only in respect of organic fertiliser.
Beveridge, writing in 1911[2], noted that in North Uist, cockle shells were still
being burned within a layer of peats to make lime. Peat was also used to make
lime from the abundant shells "at the South end of the Broad Bay" in an
operation integrated with the Lewis Chemical Works (see below). In more
recent times, the use of shell sand in reseeding schemes during the 1960s may
have involved more than 100,000 tonnes, while all but one of more than a
thousand reseeds assisted by the Western Isles Integrated Development Programme of the 1980s used shell sand[1].

THE MACAULAY INSTITUTE

The Western Isles can justly claim some credit for the founding of the Macaulay Institute for Soil Research in 1930. Dr. T.B.Macaulay, whose ancestors came from Uig in Lewis, became President of the Sun Life Insurance Company of Canada. During travels to the Highlands and Islands in the 1920s, he became so depressed by the agricultural circumstances of the area that he decided to do something about it. He consulted Dr. Robert Greig of the Department of Agriculture for Scotland and Dr. William Ogg of the East of Scotland College of Agriculture, and between them they came up with the idea for a soil research institute, which came into being in 1930, under the directorship of Ogg. Macaulay also provided funds for the establishment of an experimental farm on peatland a few miles south of Stornoway in 1928, in an attempt to demonstrate how the agriculture of Lewis could be improved[18]. Though no longer an agricultural unit, the land is still known as Macaulay Farm; the results of the trials were written up by Ogg & Macleod[23-28]. According to a report in the *Stornoway Gazette* of 15th August 1958, when the endowment funds had been exhausted, the Department of Agriculture was asked to "carry the experiments to a conclusion" but the request was refused, and the land was allowed to revert. The work at Macaulay Farm was, however, credited with the development of the method of surface reseeding subsequently used widely in the Western Isles.

If nature at times seems to have been unkind to the people of the Western Isles, where the rocks, soils and climate seem to make life difficult, it could have been worse. It is indeed fortunate that there are ample supplies of seaweed to be had near the machair soils to enable them to be cultivated, while inland the peat which makes inland cultivation almost impossible has supplied fuel for many generations in islands with few trees or alternative sources of energy. Having said this, not all communities had easy access to peat banks, and historic records such as the Royal Commission evidence give many examples of the great distances from which peat was taken, usually by boat.

Peatland plants, like all green plants, obtain their energy from sunlight by the process of photosynthesis (the essentials of which are explained in every elementary biology textbook). Energy which is not subsequently lost by respiration is stored by the plant. This stored energy is usually released from dead plants by microbial decay but such decay is very limited in peat bogs, and peat accordingly constitutes an energy store.

According to the *Orkneyinga Saga*, Turf-Einar, one of the Earls of Orkney, was the first man to dig peat for fuel in Scotland, obtaining it from "Torfness" (thought to be either Tarbet Ness or Burghead), "firewood being very scarce in the islands". Why Turf-Einar imported peat all the way from Tarbet Ness is not immediately clear as Orkney, like Lewis, is not noted for its lack of peat.

During the time of Sir James Matheson, oil was distilled from peat at the Lewis Chemical Works. The fascinating story of this venture is told in a manuscript of 1895 by D.Morison, reproduced in a volume on Scottish Industrial History published by the Scottish History Society in 1978[31]. Henry Caunter set up his distillation plant "at the side of a Fish Pond in the Vicinity of the Lewis Castle, where soon the descovery was made that the distilation of peat was *deadley* to Fish in a Pond." Interestingly, during their research on other Peat Works in Germany, they found that the condensers in the distillation plant had been made from empty Stornoway herring barrels! Following experiments, works began with an improved plant in 1860, and shortly after operations commenced, "a Fearful explosion took place", without injury. A rebuilt plant was back in operation by August 1861, and almost immediately terrified the workers by emitting a huge jet of flame as gases ignited, so further revisions in design delayed work till early 1862. In an area deriving a considerable portion of its income from fishing, it is interesting to note that Morison records "Town and Country Complaining of the sickning and offensive smell from the Works". Incredibly, the Lewis Chemical Works was spread over three sites, from Garrabost to the Creed, described by Morison as "An Enormous Reckless Blunder". Peat was taken from the banks of the River Creed, and the neat cuttings may be seen to this day, just to the west of the main road.

The Lewis Chemical Works ultimately produced 8.5–9 tons of Peat Tar each week. The pitch was held to be superior to coal tar for use in felt roofing, while the "Tar after the Pitch was removed" was used as a lubricant. The tar was used as an anti-foulant on ships' hulls. The plant became much more successful than its counterparts in Ireland and Germany, but mismanagement on an almost unbelievable scale, which included the addition of water to the lubricant, led to the closure of the works in 1874.

When Leverhulme acquired Lewis, he experimented in the use of peat for building houses, using peat pise walls on a timber framework, but encountered severe problems with damp. On his move to Harris, he tried the technique once more, but again failed[32].

CLIMATE, SOILS, AND AGRICULTURE

The agricultural use of land is determined by: climate, topography, wetness, and a number of soil factors including depth, slope, acidity, and susceptibility to drought[16].

The relationship between soils and climate is complex: the Macaulay Institute for Soil Research has published a map of *Land Capability for Agriculture in Scotland*[3] which indicates the climatic constraints on agriculture. This map (Fig 8.6) shows most of the coast of Lewis and the western seaboard of the Uists

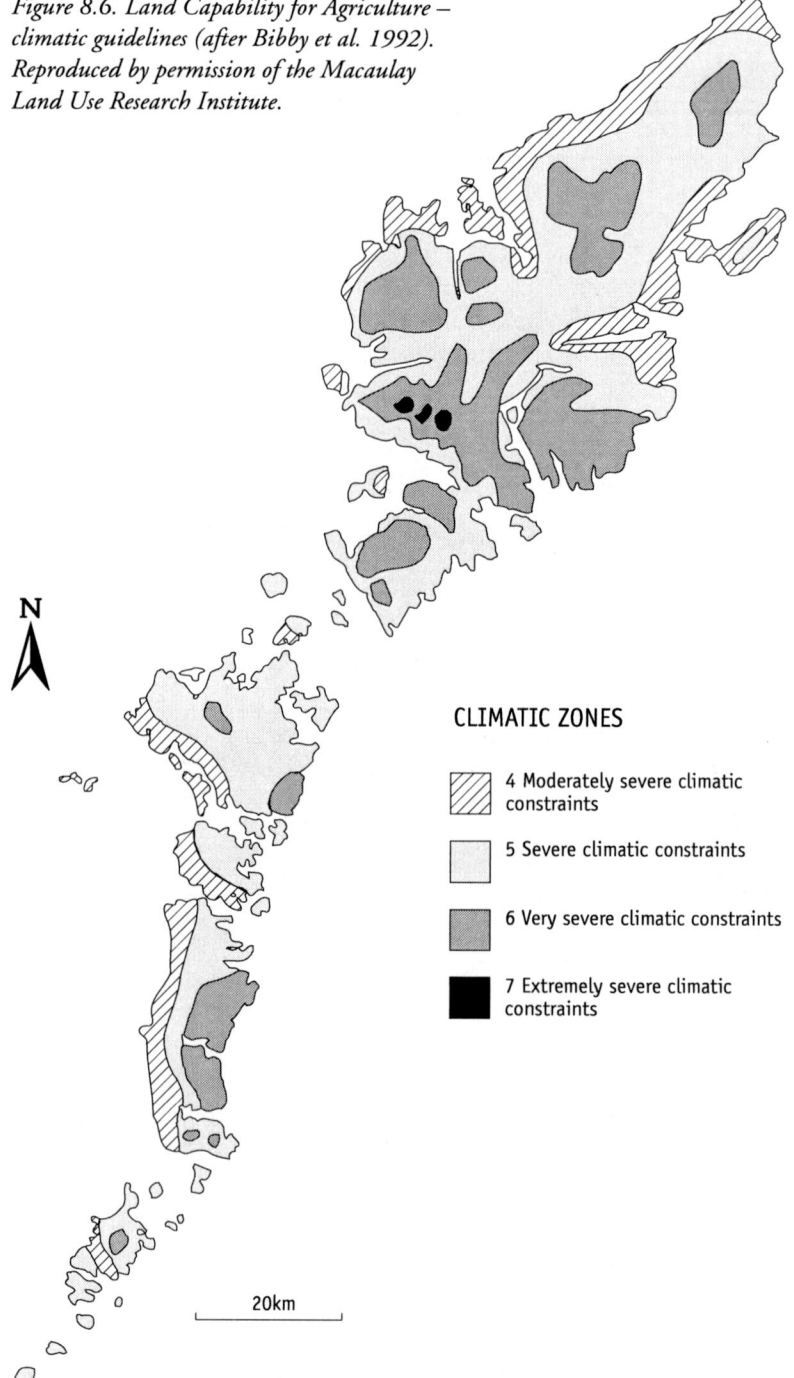

Figure 8.6. Land Capability for Agriculture –
climatic guidelines (after Bibby et al. 1992).
Reproduced by permission of the Macaulay
Land Use Research Institute.

N

CLIMATIC ZONES

4 Moderately severe climatic
constraints

5 Severe climatic constraints

6 Very severe climatic constraints

7 Extremely severe climatic
constraints

20km

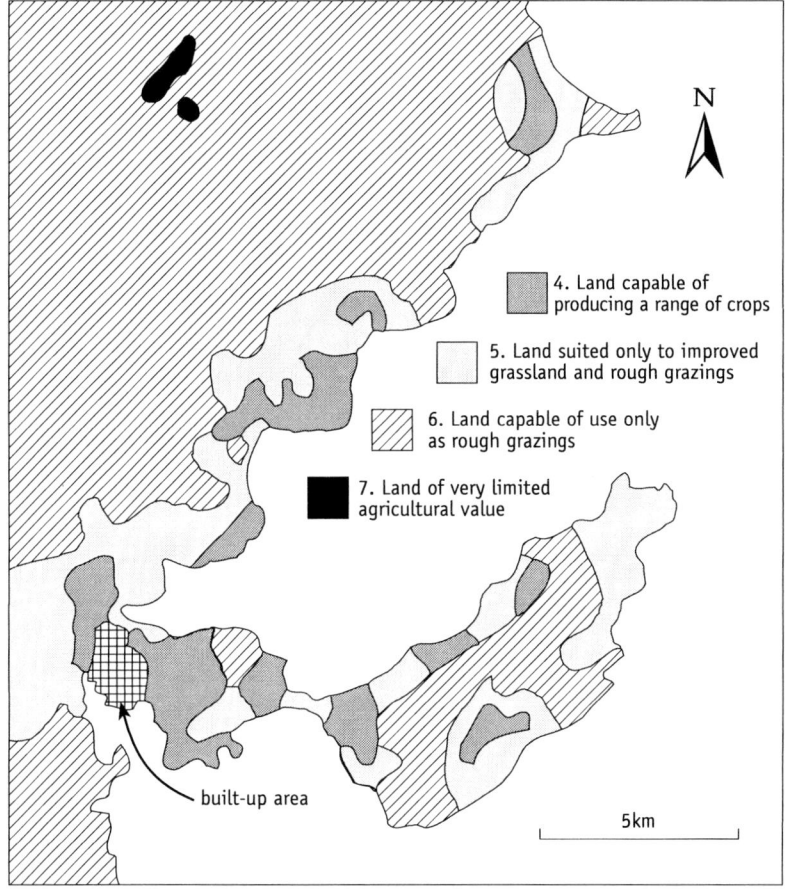

4. Land capable of producing a range of crops

5. Land suited only to improved grassland and rough grazings

6. Land capable of use only as rough grazings

7. Land of very limited agricultural value

built-up area

5km

Figure 8.7. Land Capability for Agriculture – the area around Broad Bay. Compare with Fig. 8.6. Reproduced by permission of the Macaulay Land Use Research Institute.

as having the potential for Land Class 4, the highest agricultural land class possible in the Outer Hebrides. In theory, the accumulated temperature and potential soils water deficit (see Chapter 7) around Stornoway and in Benbecula would allow the lower end of Class 3, but high exposure pushes this into Class 4[16]. The key to the Land Capability map stresses that the actual classification will be limited by soil and site limitations. The detailed Land Capability for Agriculture map for the Outer Hebrides[16] shows that Land Class 4 is very limited indeed, with most of the coast Class 5 or even 6 (Fig. 8.7). The reason for this difference is largely the interaction between the parent material and climate, encouraging the development of soils with impeded drainage and/or an organic surface layer.

Land Class 4 is "capable of producing a narrow range of crops, with enterprises based primarily on grassland with short arable breaks (e.g. barley, oats, forage crops)". Yields of arable crops are variable and, though grass can give a high yield, there can be "difficulties of production or utilisation". Land in this class is regarded as marginal for economic production of crops, and these would normally have to be of the type used for winter feed for livestock. As the authors stress, however, the intensive nature of good croft management possible on small areas can to some extent overcome some of these disadvantages[16].

Most of the land mass of the Western Isles is at the lower end of Land Class 6, where the vegetation "is dominated by plant communities with low grazing values" and is "unsuited to improvement by mechanised means" because of the steepness of slopes, the wetness of the soils, and the presence of large numbers of rocks.

MLURI has also produced a Land Capability for Forestry map, but the map for the Outer Hebrides takes the form of a small-scale insert, showing that the best land is in Class 6 (of marginal value for forestry) and that most of the land area is in Class 7 – unplantable. The authors acknowledge the limitations of the map, and that small areas would be viable for planting even within land mapped as Class 7, but generally speaking the peaty soils, high exposure to high winds and salt spray would indeed place severe limitations on commercial forestry.

APPENDIX

It is surprising how difficult it is to locate some of the most basic information about the Outer Hebrides. This section deals with boundaries, areas, distances, and coastal statistics.

BOUNDARIES

The Western Isles of Scotland stretch from Berneray, Barra (Barra Head) in the south, some 266km (165miles) to North Rona and Sula Sgeir in the North or, if you wish, a mere 218km (135 miles) from Barra Head to the Butt of Lewis at the northern tip of mainland Lewis. The greatest width is in southern Lewis – 48.5km (30 miles). Westwards, the group extends to Rockall which, like St Kilda, lies within the parish of Harris. Ignoring the Shiant Islands, the minimum separation from mainland Scotland is between Kebock Head and Rubha Reidh – 38km (23.5 miles), and the minimum distance between Harris and Skye – Finsbay to Waternish Point – is 24km (15 miles). St Kilda is about 64km (40 miles) WNW of North Uist, while North Rona is about 71km (44 miles) NNE of the Butt of Lewis. The Flannan Isles are 33km (21 miles) NW of Gallan Head, while Rockall lies 370km (230 miles) west of North Uist.

In terms of latitude and longitude, the islands stretch from 56°48.5'N (Barra Head) to 58°31.3'N (Butt of Lewis), and from 6°8.5'W (Tiumpan Head) to 7°40'W (west coast of Mingulay). North Rona lies at 59°7'N 5°49'W and Rockall at 57°36'20"N, 13°41'32"W.

The whole of the Outer Hebrides constitute the Western Isles Islands Area, administered by Comhairle nan Eilean, the Western Isles Council. Until the Comhairle was formed in 1975 by local government re-organisation, Lewis was part of Ross and Cromarty, while Harris and the Uists formed part of Inverness-shire.

An area of some 3,000 hectares around the boundary between Lewis and Harris was disputed in an action raised by Hume MacLeod against Lord Seaforth in 1804, but the boundary was not finally determined until 1853 (by which time the proprietors were the Earl of Dunmore and Sir James Matheson), when it was fixed by a Judicial Referee on behalf of the Dean of the Faculty of Advocates₁. The boundary no longer features on Ordnance Survey maps but, like the parish boundaries of Lewis, can be found on the old One-inch maps. The boundary runs along Loch Resort to Kinlochresort, then follows the Abhainn a'Chlair Bhig to the centre of Loch Chleisteir, thence via a stream, across the *bealach* between Stulabhal and Creag na Lubaig to Tom Ruisg. It then follows the SW ridge of Mullach a'Ruisg, continuing on a similar bearing to the summit of Mullach Vigadale, from which it turns roughly ESE in a straight line across a featureless slope to intercept the Abhainn a'Mhuil which it follows to the sea at Aline. Seaforth Island is wholly within both Harris and Lewis, which must have proved difficult for statisticians, though it has been claimed locally that "both counties ignored it equally".

The lucrative trade in kelp in the late 18ᵗʰ and early 19ᵗʰ centuries (see Chapter 5) meant that sources of seaweed became very valuable. A dispute between North Uist and Harris over the seaweed rights on the tiny island of Grianam (NF912778) ended up in the Court of Session and, in 1770, the Court found in favour of Macleod of Harris, despite the fact that Grianam almost touches the coast of North Uist. An appeal in 1772 confirmed the decision, but it then went to the House of Lords, who dismissed the appeal in 1781, thereby fixing a boundary between North Uist and Harris which gives Harris almost all the islands in the Sound – even those very close to the Uist coast. A paper by Alick Morrison in the *Transactions of the Gaelic Society of Inverness*₃ gives a graphic account of the boundary (modern place-names added in square brackets):

> The march enters from the waters of the Minch into the Sound of Harris through the channel between Hermatray and the lesser island of Groatay. It proceeds by the Caolas Mor between the high island of Haay [Taghay] and the dwarfish isle of Trollaman on the Uist side. It then makes a beeline for the rocks of Rangas passing straight between them. Veering perceptibly to North Uist, it passes the point of Ard Tharmaid and the Black Point of Fidday where the Macleods of Berneray used to have peat rights from MacDonald of Sleat. It proceeds through the Sound dividing Grianam Mor from Grianam Beag, which is a mere rock near the Uist coast. Its further course lies between the point of Oddernish in North Uist and the low heath-covered island of Torogay which belonged to Harris. Thus it passes down the Kyle ... enters the broad waters of the Atlantic Ocean between the historic island of Boreray and Rubh Bhoisinis in Berneray.

The 1:10,000 Ordnance Survey map confirms that Spuir, which lies between Boreray (Uist) and Pabbay (Harris), belongs to North Uist.

AREA

The 1981 Census gives a total area for the Western Isles of 289,798ha, and this figure is the one cited by most official sources. The Scottish Development Department$_4$ gives a total area of 3,087km² (308,700ha), of which 189km² (18,900ha) is inland water, giving a total land area of 2,898km² (289,800ha). Intertidal areas are not included.

Parish / Island	Darling (1955), Table 10$_2$		Caird (1989)$_1$	
	Square miles	Hectares	Imperial Acres	Hectares
Barvas	152.4	39,472		
Uig	201.7	52,240		
Lochs	179.1	46,387		
Stornoway	98.7	25,563		
Lewis	631.9	163,662	404,419	163,666
Harris	193.4	50,091	123,752	50,082
North Uist	118.0	30,562	75,513	30,560
South Uist	140.8	36,467	90,093	36,460
Barra	34.7	8,987	22,222	8,993
Lewis & Harris	825.3	213,752.7	528,171	213,748
Uists & Barra	293.5	76,016	187,828	76,012.95
Western Isles	1118.8	289,769	715,999	289,761

Table 9.1. Areas of parishes and islands in the Western Isles

COASTLINE

The Scottish Development Department$_4$ gives a total coastline for the main islands of the Western Isles of 1,813km. The Joint Nature Conservation Committee's Coastal Resource Survey gives a total of 2,102km. Detailed figures are given in the table below. The JNCC Coastal Resource Survey gives a total figure for Scotland's coastline of 11,800km, so that the Outer Hebrides make up 17.8% of Scotland's coast.

High Water Mark	1	2	3	4	5	6	Total
Muddy Shore	7.0	1.5	25.5	1.0			35.0
Sandy Shore	0.50	21.5	135.0	22.5	13.0	6.0	198.5
Rocky Shore	425.5	234.5	281.0	114.5	52.5	97.5	1,205.5
Shingle Shore	233.0	53.0	167.5	10.5	7.5	8.5	480.0
Mixed Sediments	15.5	2.5	32.0	3.5		1.0	54.5
Saltmarsh	11.5	0.5	13.0	1.0	0.5		26.5
No obvious	25.5	8.5	11.0	2.0	0.5	3.0	50.5
Terrestrial							
Shingle	6.0		0.5	0.5			7.0
Sand Dune & machair	46.0	18.0	112.5	15.0	11.0	4.5	207.0
Vertical cliff >20m	67.5	15.0	2.0		2.0	10.0	96.5
Vertical cliff <20m	29.5	3.5				3.0	36.0
Non-vert cliff >20m	16.5	29.5	1.0		0.5	10.5	58.0
Non-vert cliff <20m	19.0	1.5				3.5	24.0
Artificial embankment	2.0						2.0
Other	584.0	255.0	549.0	139.5	60.5	84.5	1,672.5
Total Length							
Mean High Water	770.5	322	665	155	74.0	116.0	2,102.5
Terrestrial	770.5	322.5	665	155	74.0	116.0	2,103.0
Area							
Flats	1166.5	95.0	5551.0	40.0	367.0	24.0	7,243.5
Saltmarsh	195.0	6.0	131.0	13.0	2.0	1.0	348.0
Dune & machair	1224.0	800.0	5153.5	260.0	416.0	110.0	7,963.5
Shingle	37.0	5.0	6.0	3.0			51.0

Table 9.2. Coastal statistics for the Western Isles

Key: 1 = Lewis & Harris mainland; 2 = Lewis & Harris detached; 3 = Uists & Benbecula mainland; 4 = Uists & Benbecula detached; 5 = Barra mainland; 6 = Barra detached. All lengths in kilometres and areas in hectares. Data from Joint Nature Conservation Committee Coastal Resource Survey.

REFERENCES

CHAPTER ONE

1. BLACKIE, J.S. 1877. North Uist and Benbecula: two sonnets by Professor Blackie. *The Celtic Magazine* (1877), 188–189.
2. CARMICHAEL, A. 1884. Grazing and agrestic customs of the Outer Hebrides. Report of Her Majesty's Commissioners of Inquiry into the Crofters and Cottars in the Highlands and Islands of Scotland. Appendix A.
3. DARLING, F.F. & BOYD, J.M. 1964. *The Highlands and Islands.* Collins, London.
4. GOODRICH-FREER, A. 1902. *Outer Isles.* Constable, London.
5. MACGILLIVRAY, W. 1828. Remarks on the sands of the Outer Hebrides. *Quarterly Journal of Agriculture*, 1, 706–714.
6. MURRAY, W.H. 1973. *The islands of Western Scotland.* Eyre Methuen, London.
7. TMS CONSULTANTS. 1995. A strategic study on tourism in the Western Isles. Unpublished report to Western Isles Tourist Board, Scottish Natural Heritage, Western Isles Council and Western Isles Enterprise.

CHAPTER TWO

1. BALD, W. 1804. Plan of Harris. The property of Alexander Hume Esquire, Surveyed by William Bald, Assistant to Mr Ainslie. National Library of Scotland.
2. BERRIDGE, N.G. 1969. A summary of the mineral resources of the crofting counties. *Rep.Br.Geol.Surv.*, Vol.1, No.5.
3. BEVERIDGE, E. 1911. *North Uist: its archaeology and topography, with notes upon the early history of the Outer Hebrides.* William Brown, Edinburgh.
4. BLACK, G.P. Ed. 1977. *Outer Hebrides: localities of geological and geomorphological importance.* Nature Conservancy Council, Newbury.
5. BRITISH GEOLOGICAL SURVEY. 1981. Geological survey of Great Britain (Scotland). 1:100, 000 Solid Geology: Lewis & Harris, Uists & Barra. Ordnance Survey, Southampton.
6. BROWN, E.D. 1978. Rockall and the limits of national jurisdiction of the UK. *Marine Policy*, 2, 181–211; 2, 275–303.
7. BURTON, K.W., COHEN, A.S., O'NIONS, R.K. & O'HARA, M.J. 1994. Archaean crustal development in the Lewisian complex of northwest Scotland. *Nature*, 370, 552–555.

8. BUTLER, C.A. 1995. The distribution, style and timing of NW-directed thrusting along the Outer Hebrides Fault Zone, NW Scotland. In: *Tectonics of the British Isles: onshore and offshore geology.* Tectonics Studies Group / Petroleum Group at the University of Durham.

9. CAIRD, J.B. 1951. The Isle of Harris. *Scottish Geographical Magazine*, 67, 85–100.

10. CLIFF, R.A., GRAY, C.M. & HUHMA, H. 1983. A Sm-Nd isotopic study of the South Harris Igneous Complex, Outer Hebrides. *Contributions to Mineralogy and Petrology*, 82, 91–98.

11. COMPSTON, W., FROUDE, D.O., IRELAND, T.R., KINNY, P.D., WILLIAMS, I.S., WILLIAMS, I.R. & MYERS, J.S. 1985. The age of (a tiny part of) the Australian continent. *Nature*, 317, 559–560.

12. COWARD, M.P., FRANCIS, P.W., GRAHAM, R.H., MYERS, J.S. & WATSON, J. 1969. Remnants of an early metasedimentary assemblage in the Lewisian complex of the Outer Hebrides. *Proceedings of the Geologists' Association*, 80, 387–408.

13. DAVIES, F.B., LISLE, R.J. & WATSON, J.V. 1975. The tectonic evolution of the Lewisian complex in Northern Lewis, Outer Hebrides. *Proceedings of the Geological Association*, 86, 45–61.

14. DEARNLEY, R. 1963. The Lewisian complex of South Harris with some observations on the metamorphosed basic intrusions of the Outer Hebrides, Scotland. *Quarterly Journal of the Geological Society of London*, 119, 243–312.

15. DEARNLEY, R. & DUNNING, F.W. 1968. Metamorphosed and deformed pegmatites and basic dykes in the Lewisian Complex of the Outer Hebrides and their geological significance. *Quarterly Journal of the Geological Society of London*, 123, 353–378.

16. DONALDSON, C.H. 1983. Tertiary igneous activity in the Inner Hebrides. *Proceedings of the Royal Society of Edinburgh*, 83B, 65–81.

17. DUNCAN, A. 1995. *Hebridean island: memories of Scarp.* Tuckwell Press, East Linton.

18. DUFF, P.M.D. 1983. Economic geology. *Geology of Scotland.* (Ed. G.Y.CRAIG), 425–454. Scottish Academic Press, Edinburgh.

19. EMELEUS, C.H. 1991. Tertiary igneous activity. *Geology of Scotland* (Ed. G.Y.CRAIG), 455–502. The Geological Society, London.

20. EMELEUS, C.H. & GYOPARI, M.C. 1992. *British Tertiary volcanic Province.* Chapman & Hall, London. (Geological Conservation Review Series).

21. FETTES, D.J., MENDUM, J.R., SMITH, D.I. & WATSON, J.V. 1992. *Geology of the Outer Hebrides: memoir for 1:100, 000 (solid edition) geological sheets, Lewis and Harris, Uist and Barra (Scotland).* HMSO, London.

22. FRANCIS, P. 1970. The geology of *Whisky Galore. New Scientist*, 46, 274–276 (7 May 1970).

23. FRANCIS, P.W. & SIBSON, R.H. 1973. The Outer Hebrides thrust. *The early Precambrian of Scotland and related rocks of Greenland.* (Ed. R.G.PARK & J.TARNEY), 95–104. University of Keele.

24. FYFE, J.A., LONG, D. & EVANS, D. 1993. *British Geological Survey United Kingdom Offshore Regional Report: The geology of the Malin-Hebrides sea area.* HMSO, for the British Geological Survey, London. ISBN 0 11 884494 6.

25. GALER, S.J.G. 1994. Oldest rocks in Europe. *Nature*, 370, 505–506.

26. GEIKIE, A. 1906. *Scottish reminiscences.* James Maclehose & Sons, Glasgow.

27. GOODWIN, A.M. 1996. *Principles of Precambrian geology.* Academic Press, London.

28. HARDING, R.R., MERRIMAN, R.J. & NANCARROW, P.H.A. 1984. St Kilda: an illustrated account of the geology. *Rep.Br.Geol.Surv.*, vol.16, No.7.

29. HARDY, M.E. 1919. *A survey of the agricultural and mineral possibilities of Lewis and Harris.* Unpublished report to the Rt. Hon. Lord Leverhulme.

30. HARRIS, A.L. 1991. The growth and structure of Scotland. *Geology of Scotland* (Ed. G.Y.CRAIG), 1–24. The Geological Society, London.

31. HEADRICK, J. 1800. *Report on the Island of Lewis.* London.

32. HEDDLE, W.F. 1888. On the general and geological features of the Outer Hebrides. *The vertebrate fauna of the Outer Hebrides* (J.A. HARVIE-BROWN & T.E.BUCKLEY), 227–244. David Douglas, Edinburgh.

33. HOLMES, A. 1965. *Principles of physical geology.* Thomas Nelson & Sons, London.

34. JEHU, T.J. & CRAIG, R.H. 1923. Geology of the Outer Hebrides, Part I. The Barra Isles. *Transactions of the Royal Society of Edinburgh*, 53, 419–441.

35. JEHU, T.J. & CRAIG, R.H. 1925. Geology of the Outer Hebrides, Part II. South Uist and Eriskay. *Transactions of the Royal Society of Edinburgh*, 53, 615–641.

36. JEHU, T.J. & CRAIG, R.H. 1926. Geology of the Outer Hebrides, Part III. North Uist and Benbecula. *Transactions of the Royal Society of Edinburgh*, 54, 469–489.

37. JEHU, T.J. & CRAIG, R.H. 1927. Geology of the Outer Hebrides, Part IV. South Harris. *Transactions of the Royal Society of Edinburgh*, 55, 457–488.

38. JEHU, T.J. & CRAIG, R.H. 1934. Geology of the Outer Hebrides, Part V. North Harris and Lewis. *Transactions of the Royal Society of Edinburgh*, 57, 839–874.

39. JOHNSTONE, G.S. & MYKURA, W. 1989. *British regional geology: the Northern Highlands of Scotland.* Fourth Edition. HMSO, London.

40. LAILEY, M., STEIN, A.M. & RESTON, T.J. 1989. The Outer Hebrides fault: a major Proterozoic structure in NW Britain. *Journal of the Geological Society, London*, 146, 253–259.

41. LAMBERT, R.St.J., MYERS, J.S. & WATSON, J.V. 1970. An apparent age for a member of the Scourie dyke suite in Lewis, Outer Hebrides. *Scottish Journal of Geology*, 6, 214–220.

42. MACCULLOCH, J. 1819–1824. *A description of the Western Isles of Scotland.* 4 vols. London.

43. MACGILLIVRAY, W. 1830. Account of the series of islands usually denominated the Outer Hebrides. *Edinb.J.Nat.Geogr.Sci.*; 1, 245–250, 401–411; 2, 87–95, 160–165, 321–334.

44. MACPHERSON, J.1992. *Tales from Barra, told by The Coddy.* Birlinn, Edinburgh.

45. McKENZIE, A.J., BENJAMIN, R.E.K., JONES, E. & DANIELL, E.T. 1973. Special mineral development programme on the islands of Harris and Lewis, Outer Hebrides. Robertson Research International, Report No. 828. Unpublished report to HIDB, Inverness.

46. McKENZIE, A.J. & PATRICK, D.J. 1971. Regional mineral survey of the Isles of Lewis and Harris, Outer Hebrides. Robertson Research Co.Ltd., Report 666. Unpublished report to HIDB, Inverness.

47. MACKINNON, A. 1974. The Madadh rocks: a Tertiary olivine dolerite sill in the Outer Hebrides. *Scottish Journal of Geology*, 10, 67–70.

48. MITCHELL, C. 1995. North Uist Coast candidate Site of Special Scientific Interest. Earth Science Documentation Series, Scottish Natural Heritage, Edinburgh (unpublished).

49. MITCHELL, C. 1995. Garry-a-Siar to Balivanich candidate Site of Special Scientific Interest. Earth Science Documentation Series, Scottish Natural Heritage, Edinburgh (unpublished).

50. MORRISON, W. 1903. The geology, physical features, botany and vertebrates of the Outer Hebrides. *History of the Outer Hebrides* (W.C.MACKENZIE), 557–579. Alexander Gardner, Paisley.

51. MOULD, P. 1953. *West-over-sea*. Oliver & Boyd, Edinburgh.

52. MURCHISON, R.I. & GEIKIE, A. 1861. On the altered rocks of the Western Islands of Scotland, and the north-western and central Highlands. *Quarterly Journal of the Geological Society of London*, 17, 171–228.

53. MYERS, J.S. 1971. The late Laxfordian granite-migmatite complex of western Harris, Outer Hebrides. *Scottish Journal of Geology*, 7, 234–284.

54. NICHOLSON, K. 1990a. Mineral resources of the Western Isles. 1. Vermiculite. *Hebridean Naturalist*, 10, 17–19.

55. NICHOLSON, K. 1990b. Mineral resources of the Western Isles. 2. Alkali feldspar. *Hebridean Naturalist*, 10, 20–31.

56. PARK, R.G. 1991. The Lewisian Complex. *Geology of Scotland* (Ed. G.Y.CRAIG), 25–64. The Geological Society, London.

57. PEACH, B.N. & HORNE, J. 1930. *Chapters on the geology of Scotland*. Oxford University Press.

58. PEACH, B.N., HORNE, J., GUNN, W., CLOUGH, C.T., HINXMAN, L.W. & TEALL, J.J.H. 1907. The geological structure of the north-west Highlands of Scotland. *Mem. Geol. Surv. U.K.*

59. PENN, I.E. & MERRIMAN, R.J. 1978. Jurassic (Toarcian to ?Bajocian) ammonites of the Shiant Islands, Outer Hebrides. *Scottish Journal of Geology*, 14, 45–53.

60. RANKIN, J.J. 1985. Life on Rockall. *The Scots Law Times*, November 1, 1985, 321–325.

61. RATCLIFFE, D.A. (Ed.). 1977. *A nature conservation review*. 2 vols. Cambridge University Press.

62. ROCK, N.M.S. 1983. The Permo-Carboniferous camptonite-monchiquite dyke-suite of the Scottish Highlands and Islands: distribution, field and petrological aspects. *Rep.Br.Geol.Surv.*, 14, No.14.

63. ROGERS, G. & PANKHURST, R.J. 1993. Unravelling dates through the ages: geochronology of the Scottish metamorphic complexes. *Journal of the Geological Society of London*, 150, 447–464.

64. SMITH, C.G. & FLOYD, J.D. 1989. Scottish Highlands and Southern Uplands mineral portfolio: hard rock aggregate resources. *British Geological Survey Technical Report* WF/89/4.

65. SMITH, D.I. & FETTES, D.J. 1979. The geological framework of the Outer Hebrides. *Proceedings of the Royal Society of Edinburgh*, 77B, 75–83.

66. SMYTHE, D.K. 1987. Deep seismic reflection profiling of the Lewisian foreland. PARK, R.G. & TERNEY, J. (Eds). Evolution of the Lewisian and comparable Precambrian high grade terrains. *Special Publication of the Geological Society of London*, No.27, 193–203.

67. SMYTHE, D.K., SOWERBUTTS, W.T.C., BACON, M. & McQUILLIN, R. 1987. Deep seismic reflection profiling of the Lewisian foreland. In: PARK, R.G. & TARNEY, J (Eds.). Evolution of the Lewisian and comparable Precambrian high grade terrains, *Special Publication of the Geological Society of London*, No.27, 683–697.

68. STEELE, R.J. & WILSON, A.C. 1975. Sedimentation and tectonism (?Permo-Triassic) on the margin of the North Minch Basin, Lewis. *Journal of the Geological Society of London*, 131, 183–202.

69. STEVENS, A. 1914. Notes on the geology of the Stornoway District of Lewis. *Transactions of the Geological Society of Glasgow*, 15, 51–63.

70. STEWART, F.H. 1965. Tertiary igneous activity. *Geology of Scotland*. (Ed. G.Y.CRAIG), 417–465. Oliver & Boyd, Edinburgh.

71. STOKER, M.S., HITCHEN, K. & GRAHAM, C.C. 1993. *The geology of the Hebrides and West Shetland shelves, and adjacent deep-water areas.* HMSO for the British Geological Survey, London. ISBN 0 11 884499 7.

72. SUTTON, J. 1973. The first three-quarters of the geological record in Britain. *The early Precambrian of Scotland and related rocks of Greenland* (Ed. R.G.PARK & J.TARNEY), 9–12. University of Keele.

73. SUTTON, J. & WATSON, J.V. 1951. The pre-Torridonian metamorphic history of the Loch Torridon and Scourie areas in the north-west Highlands, and its bearing on the chronological classification of the Lewisian. *Quarterly Journal of the Geological Society of London*, 106, 241–307.

74. TAYLOR, S.R. & MACLENNAN, S.M. 1996. The evolution of continental crust. *Scientific American*, January 1996, 60–65.

75. THOMPSON, F. 1968. *Harris and Lewis.* David & Charles, Newton Abbot.

76. WATSON, J. 1969. The Precambrian gneiss complex of Ness, Lewis, in relation to the effects of Laxfordian regeneration. *Scottish Journal of Geology*, 5, 269–285.

77. WATSON, J. 1977. The Outer Hebrides: a geological perspective. *Proceedings of the Geologists' Association*, 88, 1–14.

78. WATSON, J. 1983. Lewisian. *Geology of Scotland.* (Ed. G.Y.CRAIG), 23–47. Scottish Academic Press, Edinburgh.

79. WHITE, S.H. & GLASSER, J. 1987. The Outer Hebrides Fault Zone: evidence for normal movements. *In:* Evolution of Lewisian and comparable high grade terrains. *Geological Society of London Special Publication*, 27, 175–183.

80. WHITEHOUSE, M.J. 1993. Age of the Corodale gneisses, South Uist. *Scottish Journal of Geology*, 29, 1–7.

81. WHITTOW, J.B. 1977. *Geology and scenery in Scotland.* Penguin Books, Harmondsworth.

CHAPTER THREE

1. ANGUS, S. & ELLIOTT, M.M. 1992. Problems of erosion in Scottish machair with particular reference to the Outer Hebrides. In: CARTER, R.W.G., CURTIS, T.G.F. & SHEEHY-SKEFFINGTON, M.J. *Coastal dunes: geomorphology, ecology and management for conservation*, 93–112. A.A.Balkema, Rotterdam.

2. BADEN-POWELL, D. & ELTON, C.S. 1937. On the relation between a raised beach and an Iron Age midden on the Island of Lewis, Outer Hebrides. *Proc.Soc.Antiq.Scot.*, 71, 775–776.

3. BALLANTYNE, C.K. 1990. The Late Quaternary glacial history of the Trotternish Escarpment, Isle of Skye, Scotland, and its implications for ice-sheet reconstruction. *Geologists' Association Proceedings*, 101, 171–186.

4. BENN, D.I., LOWE, J.J. & WALKER, M.J.C. 1992. Glacier response to climatic change during the Loch Lomond Stadial and early Flandrian: geomorphological and and palynological evidence from the Isle of Skye, Scotland. *Journal of Quaternary Science*, 7, 125–144.

5. BENNETT, M.R. & BOULTON, G.S. 1993. A reinterpretation of Scottish "hummocky moraine" and its significance for the deglaciation of the Scottish Highlands during the Younger Dryas or Loch Lomond Stadial. *Geological Magazine*, 130, 301–318.

6. BEVERIDGE, G. 1926. The submerged forest and peat off Vallay, North Uist. *Scottish Naturalist*(1926), 24–25.

7. BLACK, G.P. Ed. 1977. *Outer Hebrides: localities of geological and geomorphological importance.* Nature Conservancy Council, Newbury.

8. BOULTON, G.S. 1967. The development of a complex supraglacial moraine at the margin of Sorbreen, NyFriesland, Vestspitzbergen. *Journal of Glaciology*, 6, 717–735.

9. BOULTON, G.S., PEACOCK, J.D. & SUTHERLAND, D.G. 1991. Quaternary. *Geology of Scotland* (Ed. G.Y.CRAIG), 503–543. The Geological Society, London.

10. CAMPBELL, J.F. 1873. Notes on the glacial phenomena of the Outer Hebrides. *Quarterly Journal of the Geological Society of London*, 29, 545–548.

11. CARMICHAEL, A. 1928. *Carmina Gadelica.* Volume I. Oliver & Boyd, Edinburgh.

12. CAVILL, J.E. & MERRITT, J.W. 1993. The sand and gravel resources of the Western Isles, Scotland: results of a reconnaissance survey. *British Geological Survey Technical Report* WA/93/58R.

13. COWARD, M.P. 1977. Anomalous glacial erratics in the southern part of the Outer Hebrides. *Scottish Journal of Geology*, 13, 185–188.

14. COWIE, T.G. 1983. Coastal erosion and archaeology in Lewis and Harris. *Hebridean Naturalist*, 7, 45–57.

15. DAVIES, G.L. 1956. The parish of North Uist. *Scottish Geographical Magazine*, 72, 65–80.

16. DAWSON, A.G. 1978. Former sea-level changes in the Scottish Hebrides. *Hebridean Naturalist*, 3, 16–22.

17. FLINN, D. 1978. The glaciation of the Outer Hebrides. *Geological Journal*, 13, 195–199.

18. FYFE, J.A., LONG, D. & EVANS, D. 1993. *United Kingdom offshore regional report: the geology of the Malin-Hebrides sea area.* London, HMSO for the British Geological Survey.

19. GEIKIE, J. 1873. On the glacial phenomena of the Long Island or Outer Hebrides. *Journal of the Geological Society of London*, 29, 532–545.

20. GEIKIE, J. 1878. On the glacial phenomena of the Long Island or Outer Hebrides. *Journal of the Geological Society of London*, 34, 819–870.

21. GEORGE, T.N. 1966. Geomorphic evolution in Hebridean Scotland. *Scottish Journal of Geology*, 2, 1–34.

22. GODARD, A. 1965. *Recherches de geomorphologie en Ecosse du Nord-ouest.* University of Strasbourg, Strasbourg.

23. GORDON, J.E. 1981. Ice-scoured topography and its relationships to bedrock structure and ice movement in parts of northern Scotland and West Greenland. *Geografiska Annaler*, 63A, 55–65.

24. GORDON, J.E. 1993. North-west coast of Lewis. GORDON, J.E. & SUTHERLAND, D.G. (Eds) 1993. *Geological Conservation Review Series: Quaternary of Scotland*, 414–421. Chapman & Hall, London.

25. GORDON, J.E. & SUTHERLAND, D.G. (Eds) 1993. *Geological Conservation Review Series: Quaternary of Scotland.* Chapman & Hall, London.

26. HALL, A.M. 1986. Deep weathering patterns in north-east Scotland and their geomorphological significance. *Zeitschrift für Geomorphologie*, NF30, 407–422.

27. HALL, A.M. 1991. Pre-Quaternary landscape evolution in the Scottish Highlands. *Transactions of the Royal Society of Edinburgh: Earth Sciences*, 82, 1–26.

28. HALL, A. 1996. Quaternary geomorphology of the Outer Hebrides. GILBERSTON, D, KENT, M. & GRATTAN, J. (Eds.). *The Outer Hebrides: the last 14, 000 years*, 5–12. Sheffield University Press, Sheffield.

29. HALL, A.M. 1995. Was north-west Lewis glaciated during the late Devensian? *Quaternary Newsletter*, No.76, 1–7.

30. HALSTEAD, C.A. 1979. A statistical approach to the geomorphology of the Outer Hebrides. *Proceedings of the Royal Society of Edinburgh*, 77B, 85–95.

31. HAMBREY, M. 1994. *Glacial environments*. UCL Press, London.

32. HUME, J.R. 1977. *The industrial archaeology of Scotland, 2: the Highlands and Islands*. Batsford, London.

33. JARDINE, W.G. 1982. Sea-level changes in Scotland during the last 18, 000 years. *Proceedings of the Geologists' Association*, 93, 25–41.

34. JEHU, T.J. & CRAIG, R.H. 1923. Geology of the Outer Hebrides, Part I. The Barra Isles. *Transactions of the Royal Society of Edinburgh*, 53, 419–441.

35. JEHU, T.J. & CRAIG, R.H. 1925. Geology of the Outer Hebrides, Part II. South Uist and Eriskay. *Transactions of the Royal Society of Edinburgh*, 53, 615–641.

36. JEHU, T.J. & CRAIG, R.H. 1926. Geology of the Outer Hebrides, Part III. North Uist and Benbecula. *Transactions of the Royal Society of Edinburgh*, 54, 469–489.

37. JEHU, T.J. & CRAIG, R.H. 1934. Geology of the Outer Hebrides, Part V. North Harris and Lewis. *Transactions of the Royal Society of Edinburgh*, 57, 839–874.

38. KILBURN, C., PITCHER, W.S. & SHACKLETON, R.M. 1965. the stratigraphy and origin of the Portaskaig boulder bed series (Dalradian). *Journal of Geology*, 4, 343–360.

39. LAMBECK, K. 1995. Glacial isostasy and water depths in the Late Devensian and Holocene on the Scottish Shelf west of the Hebrides. *Journal of Quaternary Science*, 10, 83–86.

40. LE COEUR, C. 1988. Later Tertiary warping and erosion in Western Scotland. *Geografiska Annaler*, 70A, 361–367.

41. LE COEUR, C. 1989. La question des altérites profondes dans la région des Hébrides internes (Ecosse occidentale). *Zeitschrift für Geomorphologie* N.F. Suppl.-Bd., 72, 109–124.

42. LEWIS, W.V. 1954. Pressure release and glacial erosion. *Journal of Glaciology*, 2, 417–422.

43. LINTON, D.L. 1959. Morphological contrasts between eastern and western Scotland. MILLER, R. & WATSON, J.W. (Eds.). *Essays in memory of Alan G. Ogilvie*, 16–45. Nelson, Edinburgh.

44. LINTON, D.L. 1963. The forms of glacial erosion. *Transactions of the Institute of British Geographers*, 33, 1–28.

45. MACGILLIVRAY, W. 1828. Remarks on the sands of the Outer Hebrides. *Quarterly Journal of Agriculture*, 1, 706–714.

46. MANLEY, G. 1952. *Climate and the British scene*. Collins, London.

47. MARTIN, M. 1703. *A description of the Western Isles of Scotland*. Ball, London.

48. MILNE HOME, D. 1880. Sixth report of the Boulder Committee. *Proceedings of the Royal Society of Edinburgh*, 10, 577–635.

49. PANZER, W. 1928. Zur oberflachengestalt der Ausseren Hebriden: Beobachtungen und Fragen. *Zeitschrift für Geomorphologie*, 3, 169–203.

50. PEACOCK, J.D. 1980. Glaciation of the Outer Hebrides: a reply. *Scottish Journal of Geology*, 16, 87–89.

51. PEACOCK, J.D. 1981. Quaternary Research Association excursion guide – Lewis and Harris. *Quaternary Newsletter*, 35, 45–54.

52. PEACOCK, J.D. 1984. Quaternary geology of the Outer Hebrides. BGS Reports, 16, No.2.

53. PEACOCK, J.D., AUSTIN, W.E.N., SELBY, I. GRAHAM, D.K., HARLAND, R. & WILKINSON, I.P. 1992. Late Devensian and Flandrian palaeoenvironmental changes in the Scottish continental shelf to the west of the Outer Hebrides. *Journal of Quaternary Science*, 7(2), 145–161.

54. PRICE, R.J. 1973. *Glacial and fluvioglacial landforms*. Oliver & Boyd, Edinburgh.

55. PRICE, R.J. 1983. *Scotland's environment during the last 30, 000 years*. Scottish Academic Press, Edinburgh.

56. RITCHIE, W. 1966. The post-glacial rise in sea level and coastal changes in the Uists. *Institute of British Geographers Transactions*, 39, 79–86.

57. RITCHIE, W. 1968. The coastal geomorphology of North Uist. *O'Dell Memorial Essays in Geography*, No.1. Department of Geography, University of Aberdeen.

58. RITCHIE, W. 1971. *The beaches of Barra and the Uists*. Department of Geography, University of Aberdeen.

59. RITCHIE, W. 1979. Machair development and chronology in the Uists and adjacent islands. *Proceedings of the Royal Society of Edinburgh*, 77B, 107–122.

60. RITCHIE, W. 1980. The beach, dunes and machair landforms of Pabbay, Sound of Harris. In: RANWELL, D.S. Ed. 1978. *Sand dune machair 3*, 13–19. Institute of Terrestrial Ecology, Norwich.

61. RITCHIE, W. 1985. Inter-tidal and sub-tidal organic deposits and sea level changes in the Uists, Outer Hebrides. *Scott.J.Geol.*, 21, 161–176.

62. RITCHIE, W. & MATHER, A.S. 1970. *The beaches of Lewis and Harris*. University of Aberdeen.

63. RITCHIE, W. & MATHER, A.S. 1972. Cementation in high latitude dunes. *Coastal Studies Bulletin*, 7, 95–112. H.H.Roberts Coastal Studies Institute, Louisiana State University, Technical Report No.131.

64. RUDDIMAN, W.F. & MCINTYRE, A. 1981. The North Atlantic Ocean during the last deglaciation. *Palaeogeog.Palaeoclim.Paleaoecol.*, 35, 145–214.

65. SHACKLETON, N.J., BACKMAN, J., ZIMMERMAN, H., KENT, D.V., HALL, M.A., ROBERTS, D.G., SCHNITKER, D., BALDAUF, J.G., DESPRAIRIES, A., HOMRIGHAUSEN, R., HUDDLESTUN, P., KEENE, J.B., KALTENBACK, A.J., KRUMSIEK, K.A.O., MORTON, A.C., MURRAY, J.W. & WESTBERG-SMITH, J. 1984. Oxygen isotope calibration of the onset of ice-rafting and history of glaciation in the North Atlantic region. *Nature*, 307, 620–623.

66. SISSONS, J.B. 1967. *The evolution of Scotland's scenery*. Oliver & Boyd, Edinburgh.

67. SISSONS, J.B. 1976. *The geomorphology of the British Isles: Scotland*. Methuen, London.

68. SISSONS, J.B. 1977. The Loch Lomond Readvance in the northern mainland of Scotland. *Studies in the Lateglacial environment of Scotland* (Ed. J.M.GRAY & J.J.LOWE), 44–59. Pergamon Press, Oxford.

69. SISSONS, J.B. 1980. The glaciation of the Outer Hebrides. *Scottish Journal of Geology*, 18, 81–84.

70. SISSONS, J.B. 1983. The Quaternary geomorphology of the Inner Hebrides: a review and reassessment. *Proceedings of the Geological Association*, 94, 165–175.

71. SMITH, D.I. & FETTES, D.J. 1979. The geological framework of the Outer Hebrides. *Proceedings of the Royal Society of Edinburgh*, 77B, 75–83.

72. STOKER, M.S., HITCHEN, K. & GRAHAM, C.C. 1993. *United Kingdom offshore regional report: the geology of the Hebrides and West Shetland shelves, and adjacent deep water areas*. London, HMSO, for the British Geological Survey.

73. SUGDEN, D.E. & JOHN, B.S. 1976. *Glaciers and landscape: a geomorphological approach*. Edward Arnold Ltd, London.

74. SUTHERLAND, D.G. 1984. The submerged landforms of the St Kilda archipelago, Western Scotland. *Marine Geology*, 58, 435–442.

75. SUTHERLAND, D.G. 1987. Submerged rock platforms on the continental shelf west of Sula Sgeir. *Scottish Journal of Geology*, 23, 251–260.

76. SUTHERLAND, D.G. 1993. Glen Valtos. GORDON, J.E. & SUTHERLAND, D.G. (Eds). *Quaternary of Scotland*, 425–429. Chapman & Hall, London.

77. SUTHERLAND, D.G., BALLANTYNE, C.K. & WALKER, M.J.C. 1982. A note on the Quaternary deposits and landforms of St Kilda. *Quaternary Newsletter*, 37, 1–5.

78. SUTHERLAND, D.G., BALLANTYNE, C.K. & WALKER, M.J.C. 1984 Late Quaternary glaciation and environmental changes on St Kilda, Scotland and their palaeclimatic significance. *Boreas*, 13, 261–272.

79. SUTHERLAND, D.G. & GORDON, J.E. 1993. The Quaternary in Scotland. GORDON, J.E. & SUTHERLAND, D.G. (Eds). *Quaternary of Scotland*, 13–47. Chapman & Hall, London.

80. SUTHERLAND, D.G. & WALKER, M.J.C. 1984. A late Devensian ice-free area and possible interglacial site on the Isle of Lewis, Scotland. *Nature, London*, 309, 701–703.

81. VON WEYMARN, J.A. 1974. *Coastline development in Lewis and Harris, Outer Hebrides, with particular reference to the effects of glaciation*. PhD Thesis, University of Aberdeen.

82. VON WEYMARN, J.A. 1979. A new concept of glaciation in Lewis and Harris, Outer Hebrides. *Proceedings of the Royal Society of Edinburgh*, 77B, 97–105.

83. VON WEYMARN, J.A. & EDWARDS, K.J. 1973. Interstadial site on the Island of Lewis, Scotland. *Nature, Lond.*, 246, 473–474.

84. WALKER, M.J.C. & LOWE, J.J. 1977. Postglacial environmental history of Rannoch Moor. I. Three pollen diagrams from the Kingshouse area. *J.Biogeogr.*, 4, 333–351.

85. WATSON, J. 1977. The Outer Hebrides: a geological perspective. *Proceedings of the Geological Association*, 88, 1–14.

86. WEST, R.G. 1977. *Pleistocene geology and biology with especial reference to the British Isles*. Longman, London.

CHAPTER FOUR

1. ANGUS, S. 1987. Lost woodlands of the Western Isles. *Hebridean Naturalist*, 9, 24–30.

2. ANGUS, S. 1994. The conservation importance of the machair systems of the Scottish islands, with particular reference to the Outer Hebrides. USHER, M.B. & BAXTER, J. (Ed.). *The islands of Scotland: a living marine heritage*, 95–120. HMSO, Edinburgh.

3. BALD, W. 1805. Plan of the Island of South Uist, the property of Ranald George McDonald Esq. of Clanranald. Scottish Record Office. RHP 38151.

4. BALD, W. 1805. Plan of the Island of Benbecula, the property of Ranald George McDonald Esq. of Clanranald. Scottish Record Office. RHP 3028.

5. BALD, W. 1805. Plan of the Estate of Boisdale, the property of Colonel Alexander Macdonald. Scottish Record Office. RHP 3066.

6. BASSETT, J.A. & CURTIS, T.G.F. 1985. The nature and occurrence of sand dune machair in Ireland. *Proceedings of the Royal Irish Academy*, 85B, 1–20.

7. BEVERIDGE, E. 1911. *North Uist: its archaeology and topography, with notes upon the early history of the Outer Hebrides.* William Brown, Edinburgh.Edinburgh.

8. BLACK, G.P. Ed. 1977. *Outer Hebrides: localities of geological and geomorphological importance.* Nature Conservancy Council, Newbury.

9. CAIRD, J.B. 1951. The Isle of Harris. *Scottish Geographical Magazine*, 67, 85–100.

10. CARMICHAEL, A. 1884. Grazing and agrestic customs of the Outer Hebrides. Report of Her Majesty's Commissioners of Inquiry into the Crofters and Cottars in the Highlands and Islands of Scotland. Appendix A.

11. CRAWFORD, I.A., 1963–90. Series of interim and unpublished reports on the excavations at Udal, North Uist (as cited in Ritchie, 1979).

12. CURTIS, T.G.F. 1991. The flora and vegetation of sand dunes in Ireland. In: QUIGLEY, M.B. (Ed.). 1991. *A guide to the sand dunes of Ireland.* Compiled for the 3rd Congress of the European Union for Dune Conservation and Coastal Management. Galway, 42–66.

13. DALBY, D.H. 1981. Shetland salt marshes. *Proceedings of the Royal Society of Edinburgh*, 80B, 191–202.

14. DEVILLERS, P., DEVILLERS-TERSCHUREN, J. & LEDANT, J-P. 1991. *CORINE biotopes manual: Habitats of the European Community.* Commission of the European Communities, Luxembourg.

15. DRAPER, L. 1991. *Wave climate atlas of the British Isles.* HMSO, London.

16. EVANS, J.G. 1971. Habitat change on the calcareous soils of Britain: the impact of Neolithic man. In: SIMPSON, D.D.A. Ed.1971. *Economy and settlement in Neolithic and early Bronze Age Britain and Europe*, 27–73. Leicester University Press, Leicester.

17. FARROW, G.E. 1974. On the ecology and sedimentation of Cardium shellsands and transgressive shellbanks of Traigh Mhor, Island of Barra, Outer Hebrides. *Transactions of the Royal Society of Edinburgh*, 69, 203–229.

18. GEIKIE, A. 1901. *The scenery of Scotland.* Macmillan, London.

19. GILBERTSON, D.D., KENT, M., SCHWENNINGER, J-L., WATHERN, P.A., WEAVER, R. & BRAYSHAY, B.A. 1995. The machair vegetation of South Uist and Barra in the Outer Hebrides of Scotland: its interacting ecological, geomorphic and historical dimensions. *Ecological relations in historical times* (Ed R. BUTLIN & N. ROBERTS), 17–44. Blackwell, London.

20. GILBERTSON, D., GRATTAN, J. & SCHWENNINGER, J-L. 1996. A stratigraphic survey of the Holcene coastal dune and machair sequences. GILBERTSON, D, KENT, M. & GRATTAN, J. (Eds.). *The Outer Hebrides: the last 14, 000 years.*, 72–101. Sheffield University Press, Sheffield.

21. GRANT, J.S. 1990. Golf is such a sacred pursuit. *Stornoway Golf Club Centenary Magazine*, 21–23.

22. GREGORY, J.W. 1927. The fiords of the Hebrides. *Geogr.J.*, 69, 193–216.

23. GREGORY, J.W. 1927. No.1 – The sea lochs of the Outer Hebrides. *Trans.Geol.Soc.Glasg.*, 18, 27–39.

24. HIGHLANDS AND ISLANDS DEVELOPMENT BOARD. 1967–72. Annual Reports.

25. HOLMES, A. 1965. *Principles of physical geology.* Thomas Nelson & Sons, London.

26. IRVING, R.A. 1997. The sea bed. In: BARNE, J.H., ROBSON, C.F., KAZNOWSKA, S.S., DOODY, J.P., DAVIDSON, N.C. & BUCK, A.L. (Eds.). *Coasts and seas of the United Kingdom. regions 15 & 16 North-west Scotland: the Western Isles and west Highland.* Joint Nature Conservation Committee, Peterborough.

27. JEHU, T.J. & CRAIG, R.H. 1934. Geology of the Outer Hebrides, Part V. North Harris and Lewis. *Transactions of the Royal Society of Edinburgh*, 57, 839–874.

28. JOINT NATURE CONSERVATION COMMITTEE. 1995. *Council Directive on the Conservation of natural habitats and wild fauna and flora (92/43/EEC) – the Habitats Directive: a list of possible Special Areas of Conservation in the UK: list for consultation.* Prepared by JNCC on behalf of English Nature, Countryside Council for Wales and Scottish Natural Heritage.

29. JOLLY, W. 1883. The nearer St Kilda: impressions of the island of Mingulay. *Good Words*, 1883, 716–120.

30. LAWSON, B. 1997. *The Isle of Taransay.* Bill Lawson Publications, Northton.

31. MACGREGOR, A.A. 1931. *Last voyage to St Kilda.* Cassell and Co., London.

32. MACGREGOR, N.A. 1990. History of the Club. *Stornoway Golf Club Centenary Magazine*, 33–43.

33. MACLEOD, A. 1831. Account of the drainage of Lochmore and Loch Scolpig, and of the embankment at Garridu, in the Island of North Uist. *Transactions of the Highland Society*, 8, 103–107.

34. MACTAGGART, F. 1997. *Luskentyre Banks and Saltings SSSI (incorporating Luskentyre-Corran Seilebost GCR site).* Unpublished report to Scottish Natural Heritage in the Earth Science Site Documentation Series, SNH, Edinburgh.

35. MATE, I.D. 1991. The theoretical development of machair in the Hebrides. *Scottish Geographical Magazine*, 108, 35–38.

36. MATHER, A.S. & RITCHIE, W. 1977. *The beaches of the Highlands and Islands of Scotland.* Countryside Commission for Scotland, Battleby.

37. MOSS, M.R. & DICKINSON, G. 1979. Evolution of the dune and machair grassland surface of South Uist, Outer Hebrides, Scotland. *Environmental Conservation*, 6, 287–292.

38. OGILVIE, A.G. 1946. Land reclamation in the Outer Isles. *Scottish Geographical Magazine*, 61, 77–84.

39. REID, R. 1799. *Plan of the Island of North Uist belonging to Alexander Macdonald.* RHP 1306, Scottish Record Office.

40. RIDLEY, G. 1983. *A diver's guide to St Kilda.* Gordon Ridley, Glasgow.

41. RITCHIE, W. 1966. The post-glacial rise in sea level and coastal changes in the Uists. *Institute of British Geographers Transactions*, 39, 79–86.

42. RITCHIE, W. 1966. Report on Tràigh Mhor. In Confidence Report to British European Airways, Renfrew.

43. RITCHIE, W. 1968. The coastal geomorphology of North Uist. *O'Dell Memorial Essays in Geography*, No.1. Department of Geography, University of Aberdeen.

44. RITCHIE, W. 1971. *The beaches of Barra and the Uists.* Department of Geography, University of Aberdeen.

45. RITCHIE, W. 1972. The evolution of coastal sand dunes. *Scottish Geographical Magazine*, (1972), 19–35.

46. RITCHIE, W. 1975. The meaning and definition of machair. *Trans.Proc.Bot.Soc.Edinb.*, 42, 431–440.

47. RITCHIE, W. 1979. Machair development and chronology in the Uists and adjacent islands. *Proceedings of the Royal Society of Edinburgh*, 77B, 107–122.

48. RITCHIE, W. & MATHER, A.S. 1970. *The beaches of Lewis and Harris.* University of Aberdeen.

49. RITCHIE, W. & WHITTINGTON, G. 1994. Non-synchronous aeolian sand movements in the Uists: the evidence of the intertidal organic and sand deposits at Cladach Mór, North Uist. *Scottish Geographical Magazine*, 110, 40–46.

50. ROBSON, M. 1991. *Rona: the distant island.* Acair, Stornoway.

51. ROYAL COMMISSION. 1884. *Report on the Highlands and Islands and evidence taken by H.M. Commissioners of Inquiry into the conditions of crofters and cottars in the Highlands and Islands, 1883.* 4 vols.

52. SMITH, J. 1978. A new method of salt pan formation in intertidal marshes. *Transactions of the Botanical Society of Edinburgh*, 43, 127–130.

53. VON WEYMARN, J.A. 1974. *Coastline development in Lewis and Harris, Outer Hebrides, with particular reference to the effects of glaciation.* PhD Thesis, University of Aberdeen.

54. WEDDERSPOON, J. 1912. Shell middens of the Outer Hebrides. *Transactions of the Inverness Scientific Society and Field Club*, 7, 315–337.

55. WILSON, J.B., WATKINS, A.J. RAPSON, G.L. & BANNISTER, P. 1993. New Zealand machair vegetation. *Journal of Vegetation Science*, 4, 655–660.

CHAPTER FIVE

1. ANDERSON, J. 1785. *An account of the present state of the Hebrides.* Report to the Lords of the Treasury.

2. ANGUS, S. & ELLIOTT, M.M. 1992. Problems of erosion in Scottish machair with particular reference to the Outer Hebrides. In: CARTER, R.W.G., CURTIS, T.G.F. & SHEEHY-SKEFFINGTON, M.J. *Coastal dunes: geomorphology, ecology and management for conservation,* 93–112. A.A.Balkema, Rotterdam.

3. BALD, W. 1805. Map of Harris: The property of Alexander Hume Esquire, Surveyed by William Bald, Assistant to Mr Ainslie. William Ballantine's Lithography, Edinburgh 1829. National Library of Scotland.

4. BALD, W. 1805. Plan of the Island of Benbecula, the property of Ranald George McDonald Esq. of Clanranald. Scottish Record Office. RHP 3028.

5. BEVERIDGE, E. 1911. *North Uist: its archaeology and topography, with notes upon the early history of the Outer Hebrides.* William Brown, Edinburgh.Edinburgh.

6. BLACK, R. 1986. *Mac Mhaighstir Alasdair: the Ardnamurchan years.* The Society of West Highland & Island Historical Research, Isle of Coll, Argyll.

7. BLACKADDER, J. 1799. *Survey and evaluation of Lord Macdonald's Estates, 1799.* Photocopy held by Scottish Record Office Ref: RH2/8/24. Associated Plan is RHP1306, held at West Register House. Original held in Humberside County Record Office, County Hall, Beverley HU17 9BA.

8. BURNETT, R. 1986. *Benbecula.* The Mingulay Press, Torlum, Benbecula.

9. CAIRD, J.B. 1979. Land use in the Uists since 1800. *Proceedings of the Royal Society of Edinburgh,* 77B, 505–526.

10. CAMPBELL, J.L. & COLLINSON, F. 1966. *Hebridean folksongs: a collection of waulking songs by Donald MacCormick in Kilphedir in South Uist in the year 1893.* Clarendon Press, Oxford.

11. CRAWFORD, I.A., 1963–90. Series of interim and unpublished reports on the excavations at Udal, North Uist (as cited in Ritchie, 1979).

12. ELTON, C. 1938. Notes on the ecological and natural history of Pabbay and other islands in the Sound of Harris. *Journal of Ecology,* 26, 275–297.

13. EVANS, J.G. 1971. Habitat change on the calcareous soils of Britain: the impact of Neolithic man. In: SIMPSON, D.D.A. Ed.1971. *Economy and settlement in Neolithic and early Bronze Age Britain and Europe,* 27–73. Leicester University Press, Leicester.

14. FLEMING, J. 1839. Report by John Fleming concerning the present state of the estates of Benbecula and South Uist and proposed alterations and developments. Unpublished ms, located in Estate Office, Askernish. [The citations given here are from a typescript of the manuscript kindly supplied by Gill Maclean, Howmore: the original document cannot be located in the Estate Office].

15. GILBERTSON, D., GRATTAN, J. & SCHWENNINGER, J-L. 1996. A stratigraphic survey of the Holcene coastal dune and machair sequences. GILBERTSON, D, KENT, M. & GRATTAN, J. (Eds.). *The Outer Hebrides: the last 14, 000 years,* 72–101. Sheffield University Press, Sheffield.

16. HUNTER, J. 1976. *The making of the crofting community.* John Donald, Edinburgh.

17. LAWSON, B. 1997. *The Isle of Taransay.* Bill Lawson Publications, Northton.

18. MACDONALD, A. & MACDONALD, A. 1924. *The poems of Alexander Macdonald (Mac Mhighstir Alasdair).* Northern Counties Newspaper and Printing and Publishing Company, Inverness.

19. MACDONALD, J. 1811. *General view of the Hebrides or Western Isles of Scotland: with observations on the means of their improvement, together with a separate account of the principal islands; comprehending their resources, fisheries, manufactures, manners and agriculture.* Published for the Board of Agriculture by Sir Richard Philips, London and Silvester Doig and Andrew Stirling, Edinburgh.

20. MACFARLANE'S GEOGRAPHICAL COLLECTIONS. C1630. 2 vols, Scottish History Society, Edinburgh.

21. MACGILLIVRAY, W. 1828. Remarks on the sands of the Outer Hebrides. *Quarterly Journal of Agriculture*, 1, 706–714.

22. MACLEAN, N. 1830. *Valuation of North Uist.* Unpublished ms. Scottish Record Office GD221/111.

23. MARTIN, M. 1703. *A description of the Western Isles of Scotland.* Ball, London.

24. MATHER, A.S. 1980. Man and machair in the nineteenth century. In: RANWELL, D.S. Ed. 1978. *Sand dune machair 3*, 12–13. Institute of Terrestrial Ecology, Norwich.

25. MATHESON, W. (Ed.). 1938. *The songs of John MacCodrum, bard to Sir James Macdonald of Sleat.* Oliver & Boyd, for the Scottish Gaelic Texts Society, Edinburgh.

26. MORRISON, A. 1967. Harris Estate papers, 1724–1754. *Transactions of the Gaelic Society of Inverness*, 45, 33–96.

27. MORRISON, A. 1982. The Grianam case, 1734–1781, the kelp industry, and the Clearances in Harris, 1811–1854. *Transactions of the Gaelic Society of Inverness*, (1982), 20–89.

28. MOULD, P. 1953. *West-over-sea.* Oliver & Boyd, Edinburgh.

29. *New Statistical Account of Scotland.* 1845. Edinburgh.

30. *[Old] Statistical Account of Scotland.* 1791–1799. Edinburgh.

31. REID, R. 1799. *Plan of the Island of North Uist belonging to Alexander Macdonald.* RHP 1306, Scottish Record Office.

32. RANDALL, R.E. 1983. Management for survival – a review of the plant ecology and protection of the "machair" beaches of north-west Scotland. In: *Sandy beaches as ecosystems*, ed. by A.McClachlan & T.Erasmus, 733–740. Dr. W.Junk Publishers, The Hague.

33. RITCHIE, W. 1968. The coastal geomorphology of North Uist. *O'Dell Memorial Essays in Geography*, No.1. Department of Geography, University of Aberdeen.

34. ROYAL COMMISSION. 1884. *Report on the Highlands and Islands and evidence taken by H.M. Commissioners of Inquiry into the conditions of crofters and cottars in the Highlands and Islands, 1883.* 4 vols.

35. SHEPHERD, I.A. 1976. Preliminary results from the beaker settlement at Rosinish, Benbecula. *Brit.Archaeol.Rep.*, 33, 209–216.

36. WALKER, J. 1764. *Report on the Hebrides.* John Donald Publishers Ltd., Edinburgh. (Edited by M.Mackay, 1980).

37. WEDDERSPOON, J. 1912. Shell middens of the Outer Hebrides. *Transactions of the Inverness Scientific Society and Field Club*, 7, 315–337.

CHAPTER SIX

1. ANGUS, S. 1983. Long Island Lagomorphs. *Hebridean Naturalist*, 7, 16–22.

2. ANGUS, S. 1994. The conservation importance of the machair systems of the Scottish islands, with particular reference to the Outer Hebrides. USHER, M.B. & BAXTER, J. (Ed.). *The islands of Scotland: a living marine heritage*, 95–120. HMSO, Edinburgh.

3. ANGUS, S. & ELLIOTT, M.M. 1992. Problems of erosion in Scottish machair with particular reference to the Outer Hebrides. In: CARTER, R.W.G., CURTIS, T.G.F. & SHEEHY-SKEFFINGTON, M.J. *Coastal dunes: geomorphology, ecology and management for conservation*, 93–112. A.A.Balkema, Rotterdam.

4. BIRD, E.F. 1987. The modern prevalence of beach erosion. *Marine Pollution Bulletin*, 18, 151–157.

5. BOYD, J.M. & BOYD, I.L. 1990. *The Hebrides: a natural history.* Collins, London.

6. CARTER, R.W.G., EASTWOOD, D.A. & BRADSHAW, P. 1992. Small scale sediment removal from beaches in Northern Ireland: environmental impact, community perception and conservation management. *Aquatic Conservation: marine and freshwater ecosystems*, 2, 95–113.

7. CAVILL, J.E. & MERRITT, J.W. 1993. The sand and gravel resources of the Western Isles, Scotland: results of a reconnaissance survey. *British Geological Survey Technical Report* WA/93/58R.

8. CRAWFORD, I.A., 1963–90. Series of interim and unpublished reports on the excavations at Udal, North Uist (as cited in Ritchie, 1979).

9. CURRIE A. 1977. Butterbur (*Petasites hybridus*) as an invading species on Machair in North-west Lewis. In: RANWELL, D.S. Ed. 1977. *Sand dune machair 2.* Institute of Terrestrial Ecology, Norwich, 22–23.

10. EILEAN AN FHRAOICH ANNUAL. 1958. Short note, p.15.

11. GILBERTSON, D., GRATTAN, J. & SCHWENNINGER, J-L. 1996. A stratigraphic survey of the Holcene coastal dune and machair sequences. GILBERTSON, D, KENT, M. & GRATTAN, J. (Eds.). *The Outer Hebrides: the last 14, 000 years*, 72–101. Sheffield University Press, Sheffield.

12. HANSOM, J.D. & COMBER, D.P.M. 1996. Eoligarry SSSI documentation and management prescription. *Scottish Natural Heritage Research, Survey and Monitoring Report*, No.49.

13. HANSON, H. & LINDH, G. 1993. Coastal erosion – an escalating environmental threat. *Ambio*, 22, 188–195.

14. HARRIS, T. & RITCHIE, W. 1989a. *Dune and machair erosion in the Barvas area: a preliminary survey.* Department of Geography, University of Aberdeen. Unpublished report to Nature Conservancy Council, Inverness.

15. HARRIS, T. & RITCHIE, W. 1989b. *Dune and machair erosion in the Luskentyre area: a preliminary survey.* Department of Geography, University of Aberdeen. Unpublished report to Nature Conservancy Council, Inverness.

16. H R WALLINGFORD. 1995. *Survey of coastal erosion in the Western Isles.* Unpublished Report to Minch Project, Stornoway. Report No. EX 3155.

17. LAWSON, B. 1994. *The Teampuill on the Isle of Pabbay: a Harris church in its historical setting.* Bill Lawson Publications, Northton. ISBN 1 872598 18 8.

18. LAWSON, B. 1997. *The Isle of Taransay.* Bill Lawson Publications, Northton.

19. MACTAGGART, F. 1997. *Luskentyre Banks and Saltings SSSI (incorporating Luskentyre-Corran Seilebost GCR site).* Unpublished report to Scottish Natural Heritage in the Earth Science Site Documentation Series, SNH, Edinburgh.

20. MATHER, A.S. & RITCHIE, W. 1977. *The beaches of the Highlands and Islands of Scotland.* Countryside Commission for Scotland, Battleby.

21. MILES, A.E.W. 1989. An early Christian chapel and burial ground on the Isle of Ensay, Outer Hebrides, Scotland with a study of the skeletal remains. *BAR British Series*, 212.

22. NORTHERN ECOLOGICAL SERVICES. 1995. *Reinstatement of sand pits and quarries in coastal dune and machair systems of the Minch area.* Unpublished report to Minch Project, Stornoway.

23. RATCLIFFE, D.A. 1968. An ecological account of the bryophytes in the British Isles. *New Phytologist*, 67, 365–439.

24. RITCHIE, W. 1966. *The physiography of the machair of South Uist.* PhD Thesis, University of Glasgow.

25. RITCHIE, W. 1968. *The coastal geomorphology of North Uist.* Department of Geography, University of Aberdeen. (O'Dell Memorial Monograph No.1).

26. RITCHIE, W. 1971. *The beaches of Barra and the Uists.* Department of Geography, University of Aberdeen.

27. RITCHIE, W. 1980. The beach, dunes and machair landforms of Pabbay, Sound of Harris. In: RANWELL, D.S. Ed. 1978. *Sand dune machair 3*, 13–19. Institute of Terrestrial Ecology, Norwich.

28. RITCHIE, W. & MATHER, A.S. 1970. *The beaches of Lewis and Harris.* University of Aberdeen.

29. ROYAL COMMISSION. 1884. *Report on the Highlands and Islands and evidence taken by H.M. Commissioners of Inquiry into the conditions of crofters and cottars in the Highlands and Islands, 1883.* 3 vols.

30. SEATON, D. 1968. Bornish blow-out. *Scott.Agric.*, summer, 145–148.

31. STEPHEN, I. (Ed.). 1993. *Siud an t-Eilean – There Goes the Island.* Acair, Stornoway.

32. STORNOWAY GAZETTE. 1950. 4 August 1950.

33. WHEELER, P. 1962. In MOISLEY, H.A. (Ed.), *Uig: a Hebridean Parish. Part III.* Department of Geography, University of Glasgow.

34. Personal Communication, Bill Lawson, 1991.

35. Personal Communication, Professor William Ritchie, 1997.

36. Personal Communication, P.L.Woodworth, 1991.

CHAPTER SEVEN

1. ALLEN, S.E., CARLISLE, A. & EVANS, C.C. 1968. The plant nutrient content of rain water, . *J.Ecol.*, 56, 497–504.

2. ANGUS, S. 1991. Climate and vegetation of the Outer Hebrides. In: PANKHURST, R.J. & MULLIN, J.M. (Eds). *Flora of the Outer Hebrides*, 28–31. Natural History Museum Publications, London.

3. ANGUS, S. & ELLIOTT, M.M. 1991. Problems of erosion in Scottish machair with particular reference to the Outer Hebrides. In: CARTER, R.W.G., CURTIS, T.G.F. & SHEEHY-SKEFFINGTON, M.J. *Coastal dunes: geomorphology, ecology and management for conservation*, 93–112. A.A.Balkema, Rotterdam.

4. ARMIT, I. 1996. *The archaeology of Skye and the Western Isles.* Edinburgh University Press, Edinburgh.

5. BIBBY, J.S., DOUGLAS, H.A., THOMASSON, A.J. & ROBERTSON, J.S. 1982 *Land capability classification for agriculture.* Soil Survey of Scotland Technical Monograph. The Macaulay Institute for Soil Research, Aberdeen.

6. BIRSE, E.L. 1971. *Assessment of climatic conditions in Scotland, 3. The bioclimatic sub-regions.* The Macaulay Institute for Soil Research, Aberdeen.

7. BIRSE, E.L. 1974. The bioclimatic characteristics of Shetland. *The natural environment of Shetland* (Ed. R.GOODIER), 24–32. Nature Conservancy Council.

8. BIRSE, E.L. & DRY, F.T. 1970. *Assessment of climatic conditions in Scotland, 1: based on accumulated temperatures and potential water deficit.* The Macaulay Institute for Soil Research, Aberdeen.

9. BIRSE, E.L. & ROBERTSON, L. 1970. *Assessment of climatic conditions in Scotland, 2: based on exposure and accumulated frost.* The Macaulay Institute for Soil Research, Aberdeen.

10. BOWES, D.R. 1960. A bog burst in the Isle of Lewis. *Scottish Geographical Magazine*, 76, 21–23.

11. CAMPBELL, R.N. 1974. St Kilda and its sheep. *In: Island survivors: the ecology of the Soay sheep of St Kilda*, ed. P.A.Jewell, C.Milner & J.M.Boyd, 8–35. Athlone Press, London.

12. DUGMORE, A. & NEWTON, A. 1996. Ideas and evidence from studies of tephra. GILBERSTON, D., KENT, M. & GRATTAN, J. (Eds.). 1996. *The Outer Hebrides: the last 14, 000 years*, 45–49. Sheffield University Press, Sheffield.

13. FRENZEL, B. 1984. Mires – repositories of climatic information or self-perpetuating ecosystems. *In: Mires: swamp, bog, fen and moor. General studies*, ed. by A.J.P.Gore, 35–65. Elsevier Scientific, Amsterdam. (Ecosystems of the World 4B).

14. GILBERSTON, D.D. & GRATTAN, J.P. 1995. The environment of Tangaval. 1. Physioography and the erosional status of cultural sites at the coast. In BRANIGAN, K. & FOSTER, P. (Eds). *Barra: archaeological research on Ben Tangaval*, 5–14. Sheffield Academic Press, Sheffield.

15. GILLETT, H. 1954. Remarkable temperature variations and fog formation over snow at Stornoway, February 8, 1954. *Met.Mag.*, 83, 344–346.

16. GLOYNE, R.W. 1968. Some climatic influences affecting hill land productivity. *Hill land productivity* (edited by I.V.Hunt), pp.9–15. British Grassland Symposium No.4, Aberdeen.

17. GRANT, J.S. 1984. *The Gaelic Vikings*. James Thin, Edinburgh.

18. GRATTAN, J., CHARMAN, D. & GILBERSTON, D. 1996. The environmental impact of Icelandic volcanic eruptions: a Hebridean perspective. GILBERTSON, D, KENT, M. & GRATTAN, J. (Eds.). *The Outer Hebrides: the last 14, 000 years.*, 51–57. Sheffield University Press, Sheffield.

19. GREEN, F.W.H. 1964. The climate of Scotland. *The vegetation of Scotland* (Ed. by J.H.Burnett), pp.15–35. Oliver & Boyd, Edinburgh.

20. GREEN, F.W.H. & HARDING, R.J. 1983. Climate of the Inner Hebrides. *Proceedings of the Royal Society of Edinburgh*, 83B, 121–140.

21. GRIBBEN, J. & LAMB, H.H. 1978. Climatic change in historical times. *Climatic change*, 66–83. Cambridge Universtity Press.

22. HOLDEN, A.V. 1961. Concentration of chloride in fresh waters and rain water. *Nature*, 192, 961–962.

23. HOLDEN, A.V. 1966. A chemical study of rain and stream waters in the Scottish Highlands. *Freshwat.Salm.Fish.Res.*, No.37.

24. HUDSON, G., TOWERS, W., BIBBY, J.S. & HENDERSON, D.J. 1982. *The Outer Hebrides: Soil and land capability for agriculture*. The Macaulay Institute for Soil Research, Aberdeen.

25. LAMB, H.H. 1977. *Climate: past, present and future*. Vol.2. Climatic history and the future. Methuen & Co., London.

26. LEE, A.J. & RAMSTER, J.W., eds. 1981. *Atlas of the seas around the British Isles*. M.A.F.F., Lowestoft.

27. MAC ILLE DHUIBH, R. 1986. Seasons. Series of articles in *West Highland Free Press.*

28. MACDONALD, D. 1978. *Lewis: a history of the island*. Gordon Wright, Edinburgh.

29. MANLEY, G. 1948. The climate of the Hebrides. *New Nat.J.*, 1, 77–82.

30. MANLEY, G. 1952. *Climate and the British scene*. Collins, London.

31. MANLEY, G. 1979. The climatic environment of the Outer Hebrides. *Proc.Roy.Soc.Edinb.*, 77B, 47–59.

32. MARTIN, M. 1703. *A description of the Western Isles of Scotland*. Ball, London.

33. McVEAN, D.N. 1961. Flora and vegetation of the islands of St Kilda and North Rona in 1958. *Journal of Ecology*, 49, 39–54.

34. MCVEAN, D.N. & RATCLIFFE, D.A. 1962. *Plant communities of the Scottish Highlands*. Monographs of the Nature Conservancy, 1. H.M.S.O., London.

35. MURRAY, W.H. 1966. *The Hebrides.* Heinemann, London.

36. PEARS, N.V. 1970. Post-glacial tree lines of the Cairngorm Mountains, Scotland: some modifications based on radiocarbon dating. *Trans.Bot.Soc.Edinb.*, 40, 536–544.

37. PENNINGTON, W., HAWORTH, E.Y., BONNY, A.P. & LISHMAN, J.P. 1972. Lake sediments in northern Scotland. *Phil.Trans.Roy.Soc.*, 264B, 191–294.

38. POORE, M.E.D. & MCVEAN, D.N. 1957. A new approach to Scottish mountain vegetation. *Journal of Ecology*, 45, 401–439.

39. PRICE, R.J. 1983. *Scotland's environment during the last 30, 000 years.* Scottish Academic Press, Edinburgh.

40. RANDALL, R.E. 1973. Airborne salt deposition and its effects upon coastal plant distribution: the Monach Isles National Nature Reserve, Outer Hebrides. *Trans.Bot.Soc.Edinb.*, 42, 153–162.

41. RATCLIFFE, D.A. 1968. An ecological account of the bryophytes in the British Isles. *New Phytologist*, 67, 365–439.

42. RATCLIFFE, D.A. 1981. The vegetation. *The Cairngorms* (D.NETHERSOLE-THOMPSON & A.WATSON), 42–76. The Melven Press, Perth.

43. STORNOWAY GAZETTE. 14 February 1958

43. STORNOWAY GAZETTE. 11 July 1958.

44. STORNOWAY GAZETTE. 31 October, 1958.

45. STORNOWAY GAZETTE. April 1976.

46. WHITE, E.J. & SMITH, R.I. 1982. *Climatology maps of Great Britain.* Institute of Terrestrial Ecology, Cambridge.

CHAPTER EIGHT

1. ANGUS, S. & ELLIOTT, M.M. 1992. Problems of erosion in Scottish machair with particular reference to the Outer Hebrides. In: CARTER, R.W.G., CURTIS, T.G.F. & SHEEHY-SKEFFINGTON, M.J. *Coastal dunes: geomorphology, ecology and management for conservation*, 93–112. A.A.Balkema, Rotterdam.

2. BEVERIDGE, E. 1911. *North Uist: its archaeology and topography, with notes upon the early history of the Outer Hebrides.* William Brown, Edinburgh.Edinburgh.

3. BIBBY, J.S., DOUGLAS, H.A., THOMASSON, A.J. & ROBERTSON, J.S. 1982 *Land capability classification for agriculture.* Soil Survey of Scotland Technical Monograph. The Macaulay Institute for Soil Research, Aberdeen.

4. BLAKE, J.L. 1966. Trees in a treeless island. *Scottish Forestry*, 20, 37–47.

5. CLYMO, R.S. 1983. Peat. *In: Mires: swamp, bog, fen and moor. General studies*, ed. by A.J.P.Gore, 159–224.. Elsevier Scientific, Amsterdam. (Ecosystems of the World 4B).

6. COWIE, T.G. 1983. Coastal erosion and archaeology in Lewis and Harris. *Hebridean Naturalist*, 7, 45–57.

7. DARLING, F.F. 1955. *West Highland survey: an essay in human ecology.* Oxford University Press, Oxford.

8. DEPARTMENT OF AGRICULTURE & FISHERIES FOR SCOTLAND. 1965. *Scottish peat surveys. Vol.II. Western Highlands and Islands.* H.M.S.O., Edinburgh.

9. GILBERTSON, D., GRATTAN, J. & SCHWENNINGER, J-L. 1996. A stratigraphic survey of the Holcene coastal dune and machair sequences. GILBERTSON, D, KENT, M. & GRATTAN, J. (Eds.). *The Outer Hebrides: the last 14, 000 years,* 72–101. Sheffield University Press, Sheffield.

10. GLENTWORTH, R. 1979. Observations on the soil of the Outer Hebrides. *Proceedings of the Royal Society of Edinburgh,* 77B, 123–137.

11. GOODE, D.A. & LINDSAY, R.A. 1979. The peatland vegetation of Lewis. *Proceedings of the Royal Society of Edinburgh,* 77B, 279–293.

12. GRANT, J.W. 1979. Cereals and grass production in Lewis and the Uists. *Proceedings of the Royal Society of Edinburgh,* 77B, 527–533.

13. GWYNNE, D., MILNER, C. & HORNUNG, M. 1974. The vegetation and soils of Hirta. *In: Island survivors: the ecology of the Soay sheep of St Kilda,* ed. P.A.Jewell, C.Milner & J.M.Boyd, 36–87. Athlone Press, London.

14. HUDSON, G. 1991. Geomorphology and soils of the Outer Hebrides. PANKHURST, R.J. & MULLIN, J.M. 1991. *Flora of the Outer Hebrides,* 19–27. Natural History Museum Publications, London.

15. HUDSON, G. & HENDERSON, D.J. 1983. Soils of the Inner Hebrides. *Proceedings of the Royal Society of Edinburgh,* 83B, 107–119.

16. HUDSON, G., TOWERS, W., BIBBY, J.S. & HENDERSON, D.J. 1982. *The Outer Hebrides: Soil and land capability for agriculture.* The Macaulay Institute for Soil Research, Aberdeen.

17. LAWSON, B. 1997. *The Isle of Taransay.* Bill Lawson Publications, Northton.

18. MACAULAY INSTITUTE FOR SOIL RESEARCH. 1971. *Annual Report 1970–71.* Introduction. Macaulay Institute for Soil Research, Aberdeen.

19. MACDONALD, J. 1811. *General view of the Hebrides or Western Isles of Scotland: with observations on the means of their improvement, together with a separate account of the principal islands; comprehending their resources, fisheries, manufactures, manners and agriculture.* Published for the Board of Agriculture by Sir Richard Philips, London and Silvester Doig and Andrew Stirling, Edinburgh.

20. MACKAY, M.M. (Ed). 1980. *The Rev. Dr. Walker's Report on the Hebrides of 1764 and 1771.* John Donald, Edinburgh.

21. MOULD, P. 1953. *West-over-sea.* Oliver & Boyd, Edinburgh.

22. MURRAY, W.H. 1973. *The islands of Western Scotland.* Eyre Methuen, London.

23. OGG, W.G. & MACLEOD, A. 1930. Reclamation and cultivation of peat land in Lewis. *Scott. J.Agric.,* 23, 121–133.

24. OGG, W.G. & MACLEOD, A. 1931. Reclamation and cultivation of peat land in Lewis. – Part II. *Scott.J.Agric.,* 14, 131–140.

25. OGG, W.G. & MACLEOD, A. 1932. Reclamation and cultivation of peat land in Lewis. – Part III. *Scott.J.Agric.*, 15, 174–184.

26. OGG, W.G. & MACLEOD, A. 1933. Reclamation and cultivation of peat land in Lewis – IV. *Scott.J.Agric.*, 16, 218–229.

27. OGG, W.G. & MACLEOD, A. 1935. Reclamation and cultivation of peat land in Lewis V. *Scott.J.Agric.*, 18, 153–159.

28. OGG, W.G. & MACLEOD, A. 1937. Reclamation and cultivation of peat land in Lewis VI. *Scott.J.Agric.*, 20, 179–186.

29. PANKHURST, R.J. & MULLIN, J.M. 1991. *Flora of the Outer Hebrides.* Natural History Museum Publications, London.

30. PEARSALL, W.H. 1968. *Mountains and moorlands.* Collins, London. (Second Edition)

31. RAE, T.I. 1978. (Ed.). The beginning and the end of the Lewis Chemical Works, 1857–1874, by D.Morison. In: *Scottish industrial history: a miscellany.* Scottish History Society, Edinburgh.

32. SHIPPOBOTTOM, M.H. 1971. The first Viscount Leverhulme: a study of an architectural patron and his work. Unpublished Thesis, Manchester University, March 1971.

33. SOIL SURVEY OF SCOTLAND. 1984. *Organisation and methods: soil and land capability for agriculture 1:250 000 survey.* The Macaulay Institute for Soil Research, Aberdeen.

34. WILKINS, D.A. 1984. The Flandrian woods of Lewis (Scotland). *Journal of Ecology*, 72, 251–258.

APPENDIX

1. CAIRD, J. 1989. Early 19th Century estate plans. In: MACLEOD, F.(Ed.). *Togail tir: the map of the Western Isles*, 49–77. Acair Ltd and An Lanntair Gallery, Stornoway.

2. DARLING, F.F. 1955. *West Highland survey: an essay in human ecology.* Oxford University Press, Oxford.

3. MORRISON, A. 1982. The Grianam case, 1734–1781, the kelp industry, and the clearances in Harris. *Transactions of the Gaelic Society of Inverness*, 52, 20–89.

4. SCOTTISH DEVELOPMENT DEPARTMENT. 1990. *The Scottish environment – statistics.* Scottish Development Department, Edinburgh.

INDEX

Illustrations are denoted by bold type and definitions by italics.